Machine Learning Tutorials for Pure Mathematics and Theoretical Physics

Machine Learning Tutorials for Pure Mathematics and Theoretical Physics

editors

Andrei Constantin

University of Birmingham, UK & University of Oxford, UK

Yang-Hui He

London Institute for Mathematical Sciences, UK & University of Oxford, UK

World Scientific

NEW JERSEY • LONDON • SINGAPORE • BEIJING • SHANGHAI • HONG KONG • TAIPEI • CHENNAI • TOKYO

Published by

World Scientific Publishing Europe Ltd.
57 Shelton Street, Covent Garden, London WC2H 9HE
Head office: 5 Toh Tuck Link, Singapore 596224
USA office: 27 Warren Street, Suite 401-402, Hackensack, NJ 07601

Library of Congress Control Number: 2026017302

British Library Cataloguing-in-Publication Data
A catalogue record for this book is available from the British Library.

MACHINE LEARNING TUTORIALS FOR PURE MATHEMATICS AND THEORETICAL PHYSICS

ISBN 978-1-80729-000-9 (hardcover)
ISBN 978-1-80729-037-5 (paperback)
ISBN 978-1-80729-001-6 (ebook for institutions)
ISBN 978-1-80729-002-3 (ebook for individuals)

For any available supplementary material, please visit
https://www.worldscientific.com/worldscibooks/10.1142/Q0584#t=suppl

Desk Editors: Aanand Jayaraman/Gabriel Rawlinson

Typeset by Stallion Press
Email: enquiries@stallionpress.com

PREFACE

Machine learning has become an important tool across the sciences, including mathematics and theoretical physics. Progress in this area has been accelerated by widely recognized advances such as the application of convolutional neural networks to large-scale image classification (AlexNet, 2012), the success of reinforcement learning in strategic gameplay (AlphaGo, 2016), and the development of transformer architectures for natural language processing and generative modelling (2017).

The material in this volume is based on the research school *Machine Learning in Mathematics and Theoretical Physics*, held in the Department of Physics at the University of Oxford from 17 to 21 August 2023, and funded by the London Mathematical Society. The school was aimed primarily at PhD students and early-career researchers and consisted of lecture courses and tutorials covering applications of machine learning in algebraic geometry, differential geometry, topology, field theory, and string theory. This book follows the structure of the school and contains lecture notes as well as solved tutorial problems. The courses assume only minimal prerequisites and are intended to be accessible to a broad audience.

The chapters have been written by the lecturers and tutors of the school, whose expertise and generosity made both the event and this volume possible. We are deeply grateful to Steve A. Abel, Lara B. Anderson, Callum R. Brodie, James Gray, Thomas R. Harvey, Elli Heyes, Edward Hirst, Magdalena Larfors, Luca A. Nutricati, Yidi Qi, and Fabian Ruehle for preparing their lecture notes and tutorials and for sharing their insight into the emerging role of machine learning in mathematics and theoretical physics.

Andrei Constantin and Yang-Hui He
Oxford

CONTENTS

PART I

BASICS OF MACHINE LEARNING

The first chapter introduces the foundational architecture of neural networks — the multi-layer feed-forward network — and explains how such models are trained within the framework of supervised and semi-supervised machine learning using the backpropagation algorithm. The second chapter presents the transformer architecture (the "T" in GPT), a model that has fundamentally reshaped modern machine learning by revolutionizing natural language processing and enabling large-scale generative models.

CHAPTER 1

NEURAL NETWORKS

FABIAN RUEHLE AND CALLUM BRODIE

In the first lecture, we introduce the basics of neural networks (NNs) in Section 1.1.1 and how to train them in Section 1.1.2. Of course, there is much more to be said than what we can fit into (half) a lecture, but it will be enough to convey the mechanisms underlying essentially all NNs. Our exposition will follow Ref. [1]. In the second half, in Section 2.2, we will introduce an NN architecture known as the transformer [2] (the "T" in ChatGPT). We were torn about whether we should spend half a lecture on a single NN architecture, when there are many more conceptual things about NNs that we will have to leave out. But we think very few things transformed (pun intended) the world in the way the transformer did. It certainly exposed the public to what artificial intelligence (AI) is capable of, and we feel in a school on machine learning (ML), students should not leave without having learned about how a transformer works.

1.1 Lecture

1.1.1 *Introduction to neural networks*

For the purpose of this lecture, we want to think of NNs as a family of maps,

$$f_\theta : \mathbb{R}^{n_{\text{in}}} \to \mathbb{R}^{n_{\text{out}}}, \tag{1.1}$$

where $\theta \in \mathbb{R}^N$ are parameters that parameterize the maps f. The maps are built by concatenating

- affine transformations: $z_\mu \mapsto W_\mu^\nu z_\nu + b_\mu$ and
- element-wise nonlinear functions: $z_\mu \mapsto \sigma(z_\mu)$ alternately.

In the above, the matrix components of W are called weights and the vector components of b are called biases. We refer to all weights and biases in the NN as (trainable) parameters. Together, they form the parameters θ in (1.1). We also want to point out that we have used the Einstein summation convention, where we sum over repeated indices in mixed position (and indices $\mu, \nu, \ldots$ are raised and lowered by δ_μ^ν). As mentioned above, each affine map is followed by a nonlinear function. The fact that compositions of affine maps are affine maps necessitates weaving in nonlinearities. Without them, NNs could only ever represent affine maps, which is way too restrictive.

Often, NNs are represented by a computational graph. We give an example of a very simple NN in Figure 1.1. This architecture is known as a fully connected feed-forward NN with one hidden layer. In this example, we have $n_{\text{in}} = 2$ and $n_{\text{out}} = 1$, i.e., the NN is a family of maps from $\mathbb{R}^2 \to \mathbb{R}$. This map is a composition of an affine map from $\mathbb{R}^2$ to $\mathbb{R}^3$, followed by an element-wise application of some nonlinearity (the tanh function, say), followed by an affine map from $\mathbb{R}^3$ to $\mathbb{R}$. In this sequence of maps, the first layer ($\mathbb{R}^2$) is referred to as the input layer, the middle layer ($\mathbb{R}^3$) as a hidden layer ("hidden" because it is typically hidden from the user, who inputs the data at the input layer and sees the final result at the output layer), and the final layer ($\mathbb{R}$) is called the output layer. The name "feed-forward" indicates that the computation is carried out sequentially from left to right in the computational graph, and "fully connected" indicates

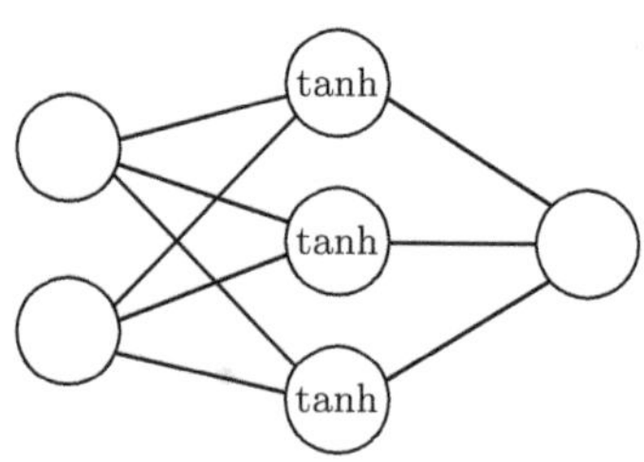

Fig. 1.1. A very simple feed-forward neural network with two input nodes, one hidden layer with three nodes, and one output node.

that the value of node v in layer $(i+1)$ depends on all nodes of the previous layer (i). The dimension, i.e., the number of hidden nodes, is called the width of an NN, while the number of hidden layers is called the depth.

Written completely explicitly, the NN in Figure 1.1 is the function

$$\begin{pmatrix} z_1 \\ z_2 \end{pmatrix} \mapsto \begin{pmatrix} w_{1,1}^{(2)} & w_{1,2}^{(2)} & w_{1,3}^{(2)} \end{pmatrix} \cdot \begin{pmatrix} \tanh(w_{1,1}^{(1)} z_1 + w_{1,2}^{(1)} z_2 + b_1^{(1)}) \\ \tanh(w_{2,1}^{(1)} z_1 + w_{2,2}^{(1)} z_2 + b_2^{(1)}) \\ \tanh(w_{3,1}^{(1)} z_1 + w_{3,2}^{(1)} z_2 + b_3^{(1)}) \end{pmatrix} + b_1^{(2)}.$$

This function is fully explicit and analytic. We can take derivatives of NNs, integrate them, etc. When people say NNs are black boxes, it is not because we do not know the function they represent (as we just mentioned, we know the analytic form of the function), it is because NNs are notoriously difficult to interpret. More precisely, this concatenation of matrix multiplication and tanh makes it extremely difficult to understand which role the parameters, i.e., the weights and biases, play in parameterizing the final map.

NNs were modelled after the human brain, where the nodes model neurons, the weights model the connection of neurons with synapses, and the activation function models a threshold at which a neuron fires, which makes the fact that a human brain has trouble interpreting them somewhat ironic. However, the idea of expanding a function in some function basis in terms of parameters is quite familiar. For example, we have Taylor expansions, which for real functions expanded around some point x_* are of the form

$$f(x) = \sum_i w_i (x - b_i),$$

where $w_i = \partial^i f(x)|_{x=x_*}$ encodes the "higher-order curvature" of f and $b_i = f(x_*)$ encodes the "starting point" about which we expand. Another example is given by Fourier expansions of the form

$$f(x) = \sum_i w_i^{(2)} \cos\left(w_i^{(1)} x + b_i^{(1)}\right),$$

where $w_i^{(2)}$ are amplitudes, $w_i^{(1)}$ are frequencies, and $b_i^{(1)}$ are phase shifts. As we can see, in these cases, the parameters are interpretable. A downside is that, for example, Taylor expansion works well if one starts around an extremum of the function but not so well if one starts at a highly curved region (in the sense that one would have to include many terms/parameters in the latter case), while Fourier expansion works well for periodic functions but needs infinitely many terms to represent a constant function, say. Of course, these expansions also become much more complicated once one tries to expand more general functions.

Still, the question is why one would use a function like an NN built from layer-wise affine maps followed by nonlinearities. A pragmatic answer would be because it works (incredibly well). To the best of my knowledge, there is no universally accepted theory for why parameterizing functions via NNs works so well. On theoretical grounds, it is known that NNs are universal function approximators, meaning that they can "learn any function" (under some mild assumptions, such as compact image of the function) in the limit where they become infinitely wide. This was proven already by Cybenko in 1989 [3] in the case of a feed-forward NN with a single hidden layer and sigmoid activation function $\sigma(x) = 1/(1+e^{-x})$. Later, this was generalized to other NNs in Refs. [4,5]; essentially, as long as the activation function is non-polynomial, NNs are universal approximators.

Since NNs can learn any function, they can in particular learn the function you are interested in. While this sounds fabulous, this also means that they can learn infinitely many functions you are not interested in, which begs again the question of why NNs work so well. Answering this question is a very active field of research; some sort of consensus seems to be that powerful-enough NNs (meaning NNs with enough parameters) tend to learn simple functions first, and simple functions tend to be the ones that generalize well to any kind of situation. The extent to which this is due to the complicated computational graph, the training of the NN with gradient descent, and other hyperparameters (we call them hyperparameters since they are chosen at the time of design, while the parameters θ of the NN are chosen randomly at the time of design and then updated during training) such as initialization that enter the problem is debated. I just want to mention one widely accepted hypothesis called the "lottery ticket hypothesis" [6], which postulates that in a big-enough NN with randomly initialized parameters, some subnet tends to be, just by chance, close to the actual function we want to compute (this subnet wins the lottery). So, the correct function is there "from the start", and all that's left to do during training is to fluctuate away/cancel/shut off the rest of the NN.

In this context, we would also like to mention a curious phenomenon known as "double descent" [7]. Typically, when one tries to fit a function with a polynomial, say, one should not simply use a polynomial of too high degree. While a polynomial of degree n can fit $n+1$ points exactly, it will also have n extrema, which means it will fluctuate wildly and hence typically not generalize well. Hence, instead of a single high-degree polynomial, one typically uses, for example, cubic splines, which are a collection of $(n-1)$ cubic functions that interpolate between n points pairwise, such that the resulting function is continuous and differentiable at the training points. On the other hand, modern NNs have hundreds of millions or even billions of parameters; nevertheless, they supposedly learn simple functions that generalize well. This fact has been observed experimentally: in the underparameterized regime (meaning the NN does not have enough parameters to even represent the function we are interested in), the error on (seen and unseen) data is still large. As the number of parameters approaches the number of parameters needed, they can be adjusted such that the error on (seen and unseen) data goes down and reaches a minimum. If one adds more parameters beyond that point and continues adjusting them, the error on seen data will continue to go down (meaning that the function fits better and better the points it is given), while the error on unseen data goes up again (meaning the NN overfits the data and learns fluctuations and noise, which prevent it from generalizing well). This phenomenon is well known as the bias–variance trade-off in statistics.

The surprise happens when one continues adding parameters until one enters the vastly overparameterized regime. At that point, the error on unseen data starts to go down again (hence the name "double descent") and can reach below the point of the first minimum (Figure 1.2). Besides vastly overparameterizing an NN, there are several other techniques that can be used to prevent overfitting, such as dropout layers, batch norm layers, early stopping, gradient or weight clipping, and weight regularization. We won't describe these in detail and refer the interested reader to Ref. [1].

Throughout this introduction, we have mentioned that the parameters are updated during training to represent the function we want to learn. Of course, this requires a notion of "target function" that we want the NN to represent. Most classically, this is the case for supervised ML. In supervised ML, we are given a sequence of input–output pairs (x_i, y_i), and the NN is supposed to learn the map from $x_i \mapsto y_i$. This could simply be a classic regression task where one tries to interpolate (or, more daringly, extrapolate) a function, but it could also be a classification task.

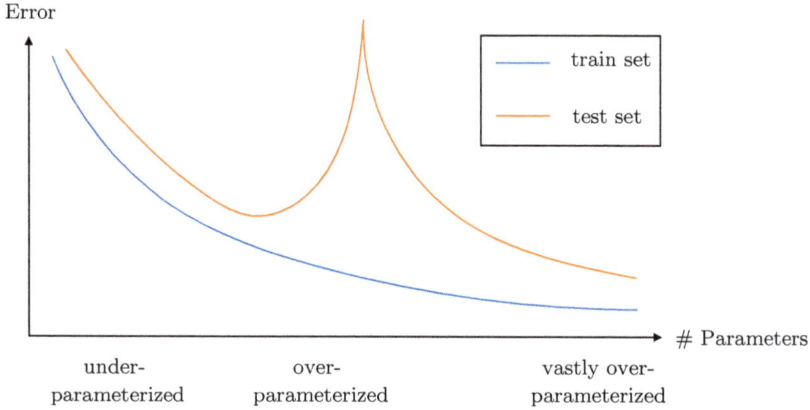

Fig. 1.2. Schematic illustration of double descent.

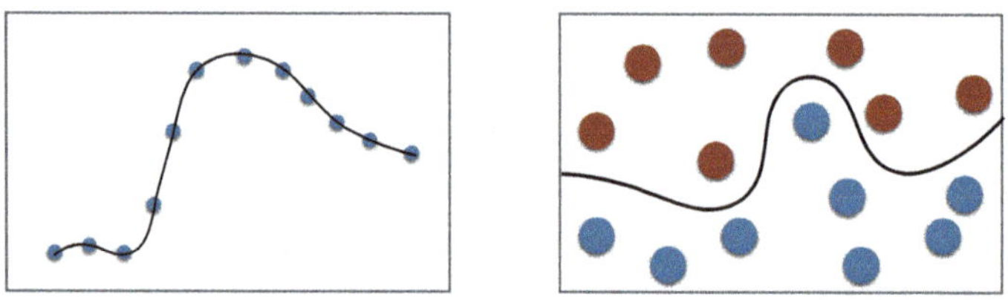

Fig. 1.3. By learning a function, an NN can be used for regression (left) or classification (right), where the function serves as a decision boundary between classes.

In the latter case, if we can learn a function that can serve as a decision boundary between two different types of objects, we can use it to tell them apart. Both cases are illustrated schematically in Figure 1.3. However, there are also other setups, such as reinforcement learning (RL), discussed in Section 7.1, which train NNs without labels.

1.1.2 *Gradient descent and backpropagation*

In order to optimize the weights and biases such that the NN gets closer to the function we want it to represent, we need a notion of "distance" from the function we want to learn, i.e., we need some sort of metric. In supervised ML, this metric is used to compute a loss, $\mathscr{L}(y_i, f_\theta(x_i))$, which indicates how close the output of the NN f_θ on a training point x_i is to the associated label y_i. The loss function is chosen such that $\mathscr{L} \geq 0$, with equality if and only if $f_\theta(x_i) = y_i$.

The type of loss function depends on the supervised ML task. For regression, popular loss functions include

$$
\begin{aligned}
&\text{Mean Absolute Error:} && \mathscr{L}_{\text{MAE}} = \frac{1}{N}\sum_{i=1}^{N} |y_i - f_\theta(x_i)|, \\
&\text{Mean Square Error:} && \mathscr{L}_{\text{MSE}} = \frac{1}{N}\sum_{i=1}^{N} (y_i - f_\theta(x_i))^2, \qquad (1.2) \\
&\text{Mean Absolute Percentage Error:} && \mathscr{L}_{\text{MAPE}} = \frac{1}{N}\sum_{i=1}^{N} |1 - f_\theta(x_i)/y_i|.
\end{aligned}
$$

Of course, MAPE requires that the labels cannot be zero. Other loss functions are also conceivable, for example the L_p norm of the distance $y_i - f_\theta(x_i)$. All of these satisfy the requirement that the loss is non-negative and vanishes if the NN has learnt to reproduce the labels y.

For a binary classification problem, the most common loss is the binary cross-entropy. Assuming the two classes are labelled by $y_i \in \{0,1\}$, the binary cross entropy loss is given by

$$\mathscr{L}_{\text{BXE}} = -\frac{1}{N} y_i \log(f_\theta(x_i)) + (1 - y_i)\log(1 - f_\theta(x_i)).$$

In this loss function, if the label is $y = 1$, only the first term contributes, and if the label is $y = 0$, only the second term contributes. Let us look at the case $y = 1$ and focus on only the first term; the discussion for $y = 0$ and the second term is analogous. If $y = 1$ and the NN output is close to 1, the loss will be very small: $\log(f_\theta) \approx \log(1) \approx 0$. If the NN output is closer to the wrong class label 0, the log becomes highly negative (and hence the loss, which has an overall minus sign, becomes very large). The loss will be non-negative if $0 \leq f_\theta \leq 1$. This is typically achieved either by using a sigmoid activation function for the output layer or by setting up two output nodes with a so-called softmax activation (for details, see Ref. [1]). Softmax activation in the final layer generalizes to multi-class classification tasks (with single-class membership) and an appropriately modified cross-entropy loss function.

Since the loss function depends on the NN output $f_\theta(x)$, it depends on the parameters θ. The task of supervised training is to update the NN parameters to minimize the loss. For modern NNs, this means finding a minimum in a multi-billion-dimensional parameter space, which means we need as fast and efficient a method as we can get. For this reason, NNs are typically trained (meaning, the NN parameters θ are updated) using a linear optimizer known as gradient descent. In gradient descent, one computes the gradient of the loss function with respect to the NN parameters, $\nabla_\theta \mathscr{L}(y, f_\theta(x))$, to find the direction in which the loss is minimized locally and updates the parameters θ proportional to this gradient:

$$\theta \to \theta - \alpha \nabla_\theta \mathscr{L}(y, f_\theta(x)). \tag{1.3}$$

This is akin to descending a mountain by looking around you to identify the steepest descent direction, taking a few steps in this direction, stopping and assessing the steepest direction from the new point, and so forth. The "number of steps" in the above example is the learning rate α in (1.3).

Since an NN is built from a concatenation of (affine and nonlinear) maps, obtaining the derivative with respect to the parameters of each layer can be done using the chain rule. We denote the result of the affine map from layer (i) to layer $(i+1)$ by

$$z_\mu^{(i+1)} = \sum_{\nu=1}^{n_i} w_{\mu\nu}^{(i+1)} \ell_\nu^{(i)} + b_\mu^{(i+1)}, \tag{1.4}$$

where $\ell_\nu^{(i)} = a^{(i)}(z_\nu^{(i)})$ is the value after applying the nonlinear activation function in layer (i). For that reason, z and ℓ are also known as pre-activation and post-activation, respectively.

Equipped with this notation, we can compute the gradient of the loss function with respect to the parameters θ, i.e., with respect to the weights w and biases b, iteratively. We first note that differentiating (1.4) gives

$$\frac{\partial z_\mu^{(i)}}{\partial w_{\mu\nu}^{(i)}} = \ell_\nu^{(i-1)}, \qquad \frac{\partial z_\mu^{(i)}}{\partial b_\nu^{(i)}} = \delta_{\mu,\nu}. \tag{1.5}$$

For the parameter derivatives of the last layer n, we thus obtain the gradient

$$\frac{\partial \mathscr{L}}{\partial \theta^{(n)}} = \frac{\partial \mathscr{L}}{\partial z_\mu^{(n)}} \frac{\partial z_\mu^{(n)}}{\partial \theta^{(n)}},$$

where we have used the chain rule. We simply computed the second term in (1.5). The first term depends on the loss function $\mathscr{L}$ but is typically a very simple expression. For example, for the MSE loss, one has from (1.2)

$$\frac{\partial \mathscr{L}}{\partial z_\mu^{(n)}} = -\frac{2}{N}\left[y_\mu - a^{(n)}(z_\mu^{(n)})\right] a'^{(n)}(z_\mu^{(n)}),$$

where $a'^{(n)}(z_\mu^{(n)})$ denotes the derivative of the last activation function with respect to its argument $z_\mu^{(n)}$. To get the dependence of the loss on the NN parameters θ in the previous layers, we can simply continue to apply the chain rule. For example (suppressing μ indices to not clutter up the equation too much),

$$\frac{\partial \mathscr{L}}{\partial \theta^{(n-1)}} = \frac{\partial \mathscr{L}}{\partial z^{(n)}} \frac{\partial z^{(n)}}{\partial z^{(n-1)}} \frac{\partial z^{(n-1)}}{\partial \theta^{(n-1)}}. \tag{1.6}$$

We have already computed the first factor on the RHS in the previous step. The last factor is given by (1.5). The second factor can be computed easily from (1.4), using that $\ell^{(i)} = a^{(i)}(z^{(i)})$:

$$\frac{\partial z^{(n)}}{\partial z^{(n-1)}} = w^{(n)}\, a'^{(n-1)}(z^{(n-1)}).$$

By continuing (1.6) backwards through layers $(n-2)$, $(n-3)$, etc., we can obtain the dependence of the loss on all NN parameters θ. For more details, see Ref. [1].

1.1.3 *Consequences of training with gradient descent*

In this section, we list a few consequences that have been observed when training an NN with gradient descent.

Special activation functions. First, note that most of the quantities in the backpropagation step, i.e., the post-activations $\ell_\nu^{(i)}$ in (1.5), have already been computed in the forward pass to get the NN output in the first place. The only thing that we need to compute in the backward pass that has not been computed in the forward pass are the derivatives of the activation function, $a'(z)$. For that reason, commonly used activation functions have derivatives that can be written in terms of the activation function itself (which has been computed in the forward pass). Common examples are

$$
\begin{aligned}
a(z) = \tanh(z) \quad &\Rightarrow \quad a'(z) = 1 - (a(z))^2, \\
a(z) = \sigma(z) = \frac{1}{1+e^{-z}} \quad &\Rightarrow \quad a'(z) = a(z)(1 - a(z)), \\
a(z) = \mathrm{ReLU}(z) = \max(0, z) \quad &\Rightarrow \quad \Theta(z).
\end{aligned}
$$

In the last equation, $\Theta(z)$ is the step function (in strict terms, for ReLU, the derivative does not contain the activation function, but it is nevertheless trivial to compute). Using any of these activation functions, the backward pass does not need to compute any new quantities which can enhance training efficiency.

Mini-batches and stochastic gradient descent. A second consequence of gradient descent is that it is easily parallelizable over the entire training set: one computes the gradient for each input–output pair separately and then simply adds the gradients. This is why GPUs or TPUs, which have thousands of cores and are designed for fast matrix multiplication, are so popular in ML. Nevertheless, one typically cannot fit the entire training set into the memory of even the largest supercomputer at once. This is especially true for vision tasks, which deal with thousands or millions of pictures, each with several megapixels, and large language models, which are trained on the essentially infinite set of texts available on the internet. To fit the data into memory, the full training set is batched into so-called mini-batches (typically of size $\mathcal{O}(1)$–$\mathcal{O}(1000)$). Then, the gradient computation is parallelized over all input–output pairs in a mini-batch and the NN parameters are updated based on the loss of this mini-batch only. Once all mini-batches have been processed, a training epoch ends. The training set is shuffled, divided into new mini-batches, and the process is repeated for another training epoch. In some LLM applications, the training set is so huge that the NN is not even trained for a single epoch (meaning you do not get through all the training data that are theoretically available), while for other applications training can last several hundreds of epochs.

The division into mini-batches and updating the parameters based on the loss of this subset rather than the entire training set means that the updates are performed based on partial information. For each mini-batch, the loss will be different, and since updates are performed after each mini-batch, gradient descent takes you in the direction of the steepest descent based on this subset only. Due to these intermittent updates, even after an entire epoch, when all training data have been processed, the point in the loss landscape where one ends up is different from the point where one would have ended up if one had performed a single gradient descent step based on the loss of the entire dataset. And since the mini-batches are shuffled after each epoch, the updates are even randomized across epochs. This is why gradient descent on mini-batches is called stochastic gradient descent. While performing updates on partial information might sound suboptimal, stochastic gradient works very well in practice. In fact, it can help overcome critical points such as local extrema or saddle points in the full landscape. At these points, the gradient would vanish, and training would stop. However, a saddle or local minima in the entire landscape is typically not a local minimum or saddle in the loss landscape of any given mini-batch, such that stochastic gradient descent can prevent getting stuck in such points. A downside is that the smaller the mini-batches, the less you can parallelize and hence the slower the training process becomes.

Vanishing/exploding gradients and the edge of chaos. As can be seen in (1.6), the gradient of layer $(i-1)$ is proportional to the gradient of layer (i). This means that gradients w.r.t. the parameters in the first few layers of the NN (those are the ones that are reached last in the error backpropagation) are susceptible to blowing up (if the gradients are larger than one) or vanishing (if the gradients are smaller than one) at a rate that is exponential in the depth of the NN. Hence, one would want the gradient of each layer to be of $\mathcal{O}(1)$ to prevent either from happening.

In order to retain finite gradients, it is worthwhile to think about what the weights and biases are actually doing. As often in physics, it is useful to consider limiting cases. Let us first study the case where the weights are huge compared to the size of the input and the biases. This means that in the pre-activation $z_\mu = w_\mu^\nu \ell_\nu + b_\mu$, all data points ℓ_ν, even those that were very close together, will be mapped very far apart, amplified by the huge weights. Behaviour where a small change in the input leads to a huge change in the output is called chaotic behaviour. In such cases, it is clear that the function changes rapidly, and the gradients will be huge. If, on the other hand, the biases are much larger than the weights and the inputs (more precisely, the weights times the post-activations of the previous layer, so the precise point where this sets in depends on the activation function used in the NN), the output of this layer becomes essentially constant (and is given by the value of the biases), and the gradients will vanish. Such a case is called ordered.

Neither the chaotic nor the ordered case is what we want, but there should be a regime in between the two phases, where gradients behave just right. The transition between the phases is actually pretty sharp [8]: the two phases are separated by a line called the edge of chaos. When initializing the weights and biases along the edge of chaos,[1] gradients (and hence parameter updates) will neither explode nor vanish. Using edge-of-chaos initialization [9], NNs with 10,000 layers have been trained successfully, which would be completely impossible if the NN is initialized in some other way. We just want to mention that there are other techniques to counter exploding or vanishing gradients, such as gradient or weight clipping, batch normalization, and skip connections which are used in UNET and ResNET architectures, but describing them here goes beyond the scope of this lecture. The interested reader can find a description in Ref. [1].

Dying ReLU problem. The fact that the gradient of the previous layer is proportional to the layer that comes after that means that activations or entire subnets can "die". Looking for example at ReLU, we see that $a(z) = a'(z) = 0$ if $z \leq 0$. This ReLU will not activate any of the other nodes it is connected to in the next layer, and since the derivative vanishes as well for $z \leq 0$, this behaviour will not change during training. This is easily combated by using a so-called "leaky ReLU", which is given by

$$a(z) = \begin{cases} z & z \geq 0, \\ az & z < 0. \end{cases}$$

Here, a is a small parameter, typically chosen to be of $\mathcal{O}(0.01)$. Now, the only point where the leaky ReLU and the derivative vanishes is the origin.

Note that the sigmoid has a similar problem (but to a lesser extent) in the sense that $\sigma(z) \sim \sigma'(z) \sim 0$ for $z \ll -1$. Unlike ReLU, neither σ nor σ' ever vanish completely, but they might become small enough to be numerically zero.

---

[1]Weights and biases are initialized randomly, for example using normal distributions. The relative size of weights and biases is then controlled by the widths of the two Gaussians, σ_w and σ_b, respectively.

1.2 Tutorial: Introduction to Neural Networks with PyTorch

This tutorial is based on a Python notebook, available at

https://github.com/andrei-const/ML_MathsPhys_Oxford2023/blob/main/oxford_ml_physmath_school_notebook_1.ipynb

which can also be run interactively using online platforms, such as Google Colab or Kaggle, or downloaded for use in Jupyter. Some familiarity with Python is assumed throughout.

Getting started with PyTorch

We begin by importing the required packages:

```
import torch
import torchvision
from torch import nn
from torch.utils.data import DataLoader
from torchvision import datasets
from torchvision.transforms import ToTensor
import matplotlib.pyplot as plt
import numpy as np
```

One of the simplest ways to build an NN in PyTorch is with `nn.Sequential`:

```
ex_seq_model = nn.Sequential(
    nn.Linear(3, 2),
    nn.ReLU(),
    nn.Linear(2, 3),
    nn.Sigmoid()
)
```

This network applies successive layers (linear maps and activation functions). Weights and biases are initialized randomly. By default, `nn.Linear` layers include bias terms, which makes them affine maps.

Q: How many layers does this example network have? How many hidden layers?

We can test the network with input tensors:

```
### COMPLETE THIS CODE CELL ###
ex_tensor = ...
ex_seq_model(ex_tensor)
```

Learning paradigms and training basics

ML tasks are broadly divided into supervised and unsupervised learning. In supervised learning, the goal is to infer a mapping from input data to known target values — classification when the outputs are discrete labels, and regression when they are continuous quantities. Training an NN in this setting involves using labelled data together with a chosen loss function and an optimization algorithm, typically based on gradient descent, to minimize prediction error. A key challenge is balancing model complexity: simple models may underfit the data (high bias), while overly flexible models may overfit (high variance). To assess a model's ability to generalize to unseen data, a separate validation set, usually comprising 10–20% of the training data, is used during training.

Q: How would training and validation losses evolve in cases of underfitting or overfitting?

During training, loss functions quantify how far the model's predictions deviate from the true targets and are used to guide optimization. A steady decrease in the validation loss generally indicates improving performance. However, the absolute value of the loss is often difficult to interpret directly. For this reason, it is common to monitor more intuitive metrics, such as accuracy in classification tasks, which measures the fraction of correctly classified examples, or, equivalently, one minus the error rate. These metrics provide a clearer sense of how well the model performs, even though they are not differentiable and thus unsuitable for use as loss functions.

Graphics processing units (GPUs) were originally developed for rendering video games, but their ability to perform a vast number of operations in parallel makes them extremely effective for accelerating many

ML tasks. In contrast, a central processing unit (CPU) executes tasks sequentially (in the single-core case) and is less efficient for large-scale parallel computations, such as those required by deep NNs. This parallelism can lead to substantial speed-ups when training on large datasets or models with many parameters. (However, not all ML applications benefit significantly from GPU acceleration.) More recently, tensor processing units (TPUs) have been developed; these correspond to specialized hardware designed specifically for ML workloads.

Getting set up to use a GPU locally can be somewhat fiddly, but cloud-hosted environments often make this straightforward. To check whether PyTorch can access a GPU or TPU on your system and to set the appropriate device, we can run the following:

```
device = (
    "cuda" if torch.cuda.is_available()
    else "mps" if torch.backends.mps.is_available()
    else "cpu"
)
print(f"Using {device} device")
```

During training, it is common to divide the dataset into smaller subsets called mini-batches. Each mini-batch is passed through the model, the loss is computed, and gradient descent is applied to update the model parameters. This process is repeated for all mini-batches, and once the entire training set has been seen, we complete one epoch. Mini-batching is often necessary when the full dataset cannot fit into memory, but it also tends to improve optimization efficiency, as the weights are updated more frequently, and the stochasticity of mini-batches helps gradient descent explore the loss landscape more effectively.

At the start of each new epoch, it is typical to shuffle the dataset before splitting it into mini-batches. Shuffling prevents the model from seeing the data in the same order every epoch, which reduces the risk of getting trapped in local minima and improves generalization.

In PyTorch, the `DataLoader` class conveniently automates batching and shuffling. This ensures that during training, the data are efficiently loaded, batched, and shuffled at the start of each epoch without manual handling. For example:

```
BATCH_SIZE = 64

train_dataloader = DataLoader(
    train_set,
    batch_size=BATCH_SIZE,
    shuffle=True
)

validation_dataloader = DataLoader(
    validation_set,
    batch_size=BATCH_SIZE,
    shuffle=True
)
```

Training and validation loops

During training, the model parameters are updated by performing gradient descent on a chosen loss function, adjusting the weights in the direction that minimizes the difference between the model's predictions and the true labels. Various refinements to basic gradient descent exist, such as variable learning rates or momentum terms, that can improve convergence and stability. These refinements correspond to different optimizers.

In PyTorch, optimizers are implemented in the `torch.optim` module. Here, we use one of the most widely adopted choices, the Adam optimizer (`torch.optim.Adam`), together with cross-entropy loss, which is standard for classification tasks.

The core training and validation loops proceed as follows. During each training epoch, we iterate over all batches in the training dataset. For each batch, we compute the model output, evaluate the loss, perform backpropagation, and update the parameters using the optimizer. After each epoch, we evaluate the updated model on the validation set to monitor its generalization performance.

In what follows, we define the functions implementing a single training and validation loop. Before each loop, we call `model.train()` or `model.eval()` to switch between training and evaluation modes, respectively. (In practice, this is only necessary if the network contains some specific types of layers, but it's good practice to always include these lines.) Gradients are reset to zero using `optimizer.zero_grad()` before backpropagation, and after computing the loss, the call to `loss.backward()` accumulates gradients for all parameters. Finally, the optimizer step `optimizer.step()` updates the parameters.

For classification tasks, the model outputs a vector of predicted probabilities across classes; the predicted label corresponds to the index of the maximum probability, obtained using `pred.argmax(1)`.

Q: Check that you understand the steps in the following training and validation loops.

```
def train_loop(dataloader, model, loss_fn, optimizer):
    size = len(dataloader.dataset)
    model.train()
    for batch, (X, y) in enumerate(dataloader):
        X, y = X.to(device), y.to(device)
        pred = model(X)
        loss = loss_fn(pred, y)
        loss.backward()
        optimizer.step()
        optimizer.zero_grad()
        if batch % 100 == 0:
            loss_value = loss.item()
            current = (batch + 1) * len(X)
            print(
                f"loss: {loss_value:>7f}  "
                f"[{current:>5d}/{size:>5d}]"
            )


def validation_loop(dataloader, model, loss_fn):
    size = len(dataloader.dataset)
    num_batches = len(dataloader)
    model.eval()
    validation_loss, correct = 0, 0
```

```
    with torch.no_grad():
        for X, y in dataloader:
            X, y = X.to(device), y.to(device)
            pred = model(X)
            validation_loss += loss_fn(pred, y).item()
            correct += (
                (pred.argmax(1) == y)
                .type(torch.float)
                .sum()
                .item()
            )
    validation_loss /= num_batches
    correct /= size
    print(
        f"Validation Error:\n"
        f" Accuracy: {(100*correct):>0.1f}%, "
        f"Avg loss: {validation_loss:>8f}\n"
    )
```

Example: MNIST digit classification

We now train a simple NN to classify handwritten digits in the MNIST dataset:

```
train_set = datasets.MNIST(
    root="data", train=True,
    download=True, transform=ToTensor()
)

validation_set = datasets.MNIST(
    root="data", train=False,
    download=True, transform=ToTensor()
)
```

Each image is a 28×28 greyscale array. The input layer size is therefore equal to the total number of pixels, the output layer size corresponds to the number of target classes (in this case, ten), and we choose an intermediate hidden layer size as follows:

```
num_in = image.numel()
num_out = len(train_set.classes)
num_hidden = 100
```

We use the following NN:

```
seq_model = nn.Sequential(
    nn.Flatten(start_dim=1),
    nn.Linear(num_in, num_hidden),
    nn.ReLU(),
    nn.Linear(num_hidden, num_out)
).to(device)
```

Training proceeds in the following way:

```
loss_fn = nn.CrossEntropyLoss()
optimizer = torch.optim.Adam(seq_model.parameters(), lr=1e-3)

epochs = 5
for t in range(epochs):
    print(f"Epoch {t+1}\n-------------------------------")
    train_loop(train_dataloader, seq_model, loss_fn,
      optimizer)
    validation_loop(validation_dataloader, seq_model, loss_fn)
print("Done!")
```

With an appropriately chosen hidden layer size and sufficient training, the model should achieve a validation accuracy above 95%, reflecting its ability to generalize from the training data to unseen examples. The specific value of the hidden dimension affects both the capacity and complexity of the network: too few hidden units may lead to underfitting, while too many can increase the risk of overfitting or unnecessary computational cost.

In this example, we used a feed-forward NN with a single hidden layer, making it a straightforward instance of deep learning. Despite its simplicity, this architecture is powerful enough to learn a high-quality representation of the handwritten digits. In more complex image-recognition problems, however, one typically employs convolutional neural networks

(CNNs). These models are designed to exploit the spatial structure of images by sharing weights across pixels.

References

[1] F. Ruehle, "Data science applications to string theory," *Phys. Rept.* **839**, 1–117, 2020.

[2] A. Vaswani, N. Shazeer, N. Parmar, J. Uszkoreit, L. Jones, A. N. Gomez, L. U. Kaiser, and I. Polosukhin, "Attention is all you need," in I. Guyon, U. V. Luxburg, S. Bengio, H. Wallach, R. Fergus, S. Vishwanathan, and R. Garnett, (eds.), *Advances in Neural Information Processing Systems*, Vol. 30. Curran Associates, Inc., 2017.

[3] G. Cybenko, "Approximations by superpositions of sigmoidal functions," *Math. Control Signals Syst.* **2**, 303, 1989.

[4] M. Leshno, V. Y. Lin, A. Pinkus, and S. Schocken, "Multilayer feedforward networks with a nonpolynomial activation function can approximate any function," *Neural Netw.* **6**, no. 6, 861–867, 1993.

[5] V. Maiorov, "Approximation by neural networks and learning theory," *J. Complex.* **22**, 102–117, 2006.

[6] J. Frankle and M. Carbin, "The lottery ticket hypothesis: Finding sparse, trainable neural networks," in *International Conference on Learning Representations*, 2019.

[7] M. Belkin, D. Hsu, S. Ma, and S. Mandal "Reconciling modern machine-learning practice and the classical bias–variance trade-off," *Proc. Natl. Acad. Sci.* **116**, no. 32, 15849–15854, 2019. https://www.pnas.org/doi/pdf/10.1073/pnas.1903070116.

[8] S. S. Schoenholz, J. Gilmer, S. Ganguli, and J. Sohl-Dickstein, "Deep information propagation," in *5th International Conference on Learning Representations, ICLR 2017, Toulon, France, April 24–26, 2017, Conference Track Proceedings*. OpenReview.net.

[9] L. Xiao, Y. Bahri, J. Sohl-Dickstein, S. Schoenholz, and J. Pennington, "Dynamical isometry and a mean field theory of CNNs: How to train 10,000-layer vanilla convolutional neural networks," in J. Dy and A. Krause (eds.), *Proceedings of the 35th International Conference on Machine Learning*. Proceedings of Machine Learning Research, Vol. 80, pp. 5393–5402. PMLR, 10–15 July 2018.

CHAPTER 2

TRANSFORMERS

FABIAN RUEHLE

2.1 Introduction

It is probably fair to say that large language models (LLMs) have transformed (pun intended) the way the public interacts actively with AI and have led to debates on what artificial general intelligence (AGI) means and whether we have already seen sparks of AGI [1]. LLMs such as ChatGPT are based on the so-called transformer architecture (this is what the "T" stands for), which was introduced together with the attention mechanism in the seminal paper "Attention is all you need" [2]. I do not know what architecture, if any, will realize AGI, but given the prominent role the transformer has played, I feel an introductory lecture on AI should explain the key ingredients in language processing and of the transformer architecture. An overview of all components is given in Figure 2.1 on the top left. We will discuss each component in turn.

2.2 Lecture

2.2.1 *Tokenization and embedding*

The first question is how to input words into a neural network (NN). The way this is typically done is by converting each word that occurs in the training set into a vector using a so-called one-hot encoding. For the sake of argument, let us say we take the Oxford American Dictionary, which contains about 300,000 words. In that case, the first word in the dictionary,

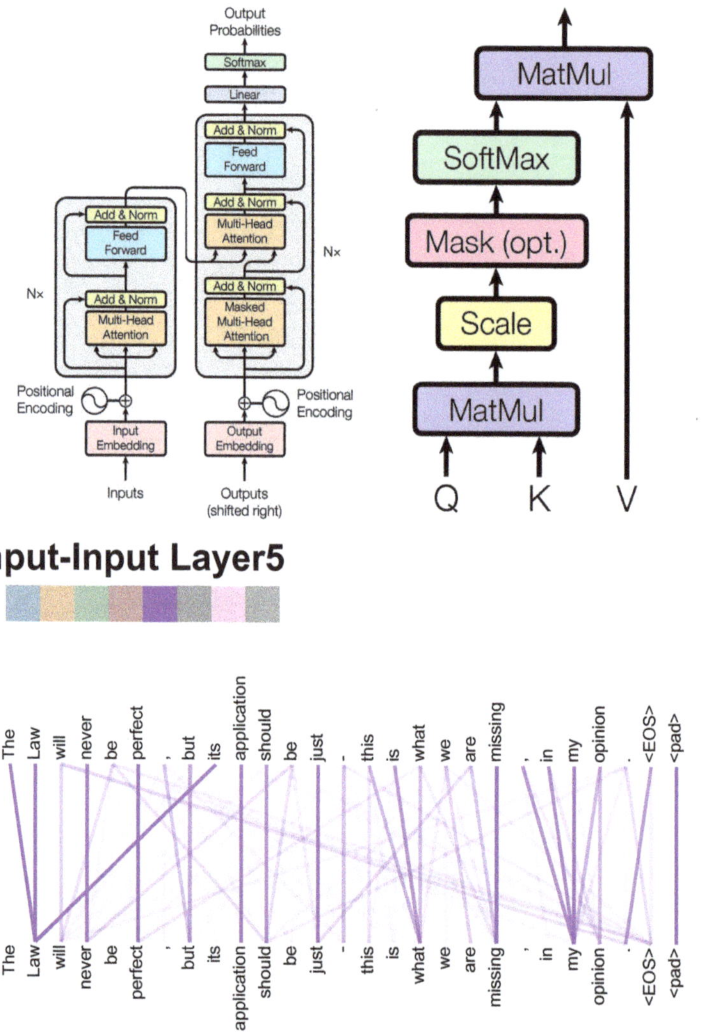

Fig. 2.1. Top left: Overview of the transformer architecture with encoder (left) and decoder (right). Top right: Attention module. Bottom: Example of attention scores in a sentence.

Source: These figures have been reproduced from Ref. [2].

"a", would be encoded by a vector of $\mathbb{R}^{300,000}$, with a single one at the first position and zeros in all others. The second word in the dictionary, "Aardvark", would be a 300k-dimensional vector with a one at the second position and zero every else and so on, until the last word "Zyzzyva".

This step is known as tokenization (sometimes, instead of full words, word parts are tokenized),

$$|\text{a}\rangle = \begin{pmatrix}1\\0\\0\\\vdots\\0\end{pmatrix}, \; |\text{Aardvak}\rangle = \begin{pmatrix}0\\1\\0\\\vdots\\0\end{pmatrix}, \quad \ldots \quad, |\text{Zyzzyva}\rangle = \begin{pmatrix}0\\1\\0\\\vdots\\0\end{pmatrix}.$$

We represent the vector corresponding to a tokenized word w by $|w\rangle$.

While this encoding allows to feed a word (or a sentence or text for that matter) to an NN, the problem with this encoding is that it is very wasteful (it consists of mostly zeros). Moreover, it is simply an embedding of words into a huge vector space that is based on some random ordering and does not capture the structure or semantics of language. It is hence unclear and highly unlikely that this representation will be beneficial to an NN. Instead of using this embedding, we can train an NN to learn a "better, more useful" embedding into a much smaller vector space. To do that, we simply multiply each 300k vector with a $512 \times 300{,}000$ matrix E with trainable parameters $e_{i,j}$,

$$|\text{a}\rangle_E = E \cdot \begin{pmatrix}1\\0\\\vdots\\0\end{pmatrix} = \begin{pmatrix}e_{1,1}\\e_{2,1}\\\vdots\\e_{512,1}\end{pmatrix}, \; \ldots \; , |\text{Zyzzyva}\rangle_E = E \cdot \begin{pmatrix}0\\0\\\vdots\\1\end{pmatrix} = \begin{pmatrix}e_{1,300000}\\e_{2,300000}\\\vdots\\e_{512,300000}\end{pmatrix},$$

where we have added a subscript E to the vectors to indicate that they are now embedded into $\mathbb{R}^{512}$ using an embedding matrix E.

This will produce an embedding into $\mathbb{R}^{512}$, and this embedding can actually be learnt in such a way that it captures the structure of language. In simple word-to-vector applications, learnt embeddings[1] have been shown to learn semantic relationships [3]. Taking the 512-dimensional vectors of various words and adding or subtracting them to obtain a new

[1]The embedding matrix E has been learnt for example by training a recurrent NN to predict the next word in a sentence.

vector which will correspond to a new word showed that the embedding represents semantics. For example,

$$|\text{queen}\rangle_E - |\text{woman}\rangle_E + |\text{man}\rangle_E = |\text{king}\rangle_E$$

or

$$|\text{Spain}\rangle_E - |\text{Madrid}\rangle_E = |\text{Germany}\rangle_E - |\text{Berlin}\rangle_E.$$

In any case, even if the embedding is not learnt separately, it is worthwhile to use a (random) matrix to embed the words into a less wasteful, smaller vector space for further processing.

2.2.2 *Positional encoding*

After the first step, we can represent a sentence with n words as a vector in $\mathbb{R}^{512}$ by concatenating the embedding vectors of the n words. In most languages, the order of words in a sentence is important. For example, the sentences "the scientist eats the chicken" and "the chicken eats the scientist" are semantically quite different. However, NN architectures such as fully connected feed-forward NNs or the transformer have no way of knowing the order of words in a sentence; all they see is a long vector in which each component is treated on equal footing. Hence, as a next step, we need some positional encoding which tells the NN about the position of the word in the sentence. In the original paper, the authors take the embedded word and add alternately a sine or cosine (whose phases are computed based on the position of the word in the sentence) to the embedding vector, but other positional encodings would work as well. Denoting the vector that indicates the position t of a word in a sentence by $|\text{PE}(t)\rangle$, we obtain a new, 512-dimensional, position-encoded vector:

$$\begin{aligned} |\text{The}\rangle_{EP} &= |\text{The}\rangle_E + |\text{PE}(0)\rangle\,, \\ |\text{scientist}\rangle_{EP} &= |\text{scientist}\rangle_E + |\text{PE}(1)\rangle\,, \ldots \end{aligned}$$

For each word, this produces a new embedding vector that has the same dimension but is overlaid with this modulated time signal to encode the word's position in the sentence. This also means that if a word (e.g., "the") appears multiple times in a sentence, it will be mapped to different vectors after this step.

2.2.3 *The transformer encoding blocks*

Given the embedded words with positional encodings added, we can start feeding them to an NN. The transformer consists of N encoder blocks (the authors choose $N = 6$ in the paper), which learn an encoding for each word that can be used for downstream tasks such as predicting the next word in a sentence or translating a sentence to another language. The encoder blocks are the grey blocks in the top left of Figure 2.1. The novelty of the transformer NN is its attention heads in each encoder block. In addition to the attention heads, the encoder blocks use skip connections, layer normalization, and traditional feed-forward NNs. We will explain these components next.

Attention heads. The transformer architecture introduces attention heads. The attention vector is computed in a two-step process. In the first step, which is illustrated in Figure 2.1 in the top right, one computes a (real-valued) score or overlap between an input word $|w_i\rangle$ and some other word $|w_j\rangle$ in an input sentence $|\sigma\rangle = |w_1\rangle \oplus \cdots \oplus |w_n\rangle$ (we obtain the vector representation of a sentence by concatenating the vector representations of all words in the sentence). Hence, the score for a sentence will be a vector in $\mathbb{R}^n$, whose jth component, the word w_j in the sentence σ, is for "understanding" the input word w_i. In the second step, we use this score value to compute an attention vector.

In more detail, the score is computed from each word vector $|w_i\rangle_{EP}$ as follows: first, we compute three new vectors for each word in the sentence called the query vector $\vec{q}(|w_i\rangle_{EP})$, key vector $\vec{k}(|w_i\rangle_{EP})$, and value vector $\vec{v}(|w_i\rangle_{EP})$.[2] They are computed by multiplying $|w_i\rangle_{EP}$ by a query matrix $\mathcal{Q}$, a key matrix $\mathcal{K}$, and a value matrix $\mathcal{V}$:

$$\vec{q}(|w_i\rangle_{EP}) = \mathcal{Q} \cdot |w_i\rangle_{EP}, \quad \vec{k}(|w_i\rangle_{EP}) = \mathcal{K} \cdot |w_i\rangle_{EP},$$
$$\vec{v}(|w_i\rangle_{EP}) = \mathcal{V} \cdot |w_i\rangle_{EP}.$$

The matrix entries in $\mathcal{Q}$, $\mathcal{K}$, and $\mathcal{V}$ are learnable parameters just like in the feed-forward case. In the paper, the authors choose these to be 64×512 matrices, such that we can think of each of them as maps from the 512-dimensional input space to a 64-dimensional output space (the query, key, and value spaces, respectively).

[2] These names are reminiscent of database queries.

We then compute the dot product (which measures the "overlap" of two vectors in the sense that it is zero for orthogonal vectors and maximized if the two vectors are the same) between the query vector of a word w_i and the key vectors of all n words w_j in the sentence. This step mimics a query to a database which would retrieve words whose key matches the query.

Given these scores, we divide them by $8 = \sqrt{64}$, where 64 is the dimension of each vector, to avoid exploding gradients, and then we take the softmax to convert the "overlaps" to "probabilities". Finally, the value vectors are weighted (i.e., multiplied) with these probabilities and summed to compute the attention score between the input word $|w_i\rangle_{PE}$ and each word $|w_j\rangle_{PE}$ in the sentence $|\sigma\rangle$.

Since we need to compute this attention vector for each input word w_i, $i = 1, \ldots, n$, we can collect the n query vectors $\vec{q}(|w_i\rangle_{EP})$ in an $n \times 64$ matrix Q, the n key vectors $\vec{k}(|w_i\rangle_{EP})$ in an $n \times 64$ matrix K, and the n value vectors $\vec{q}(|w_i\rangle_{EP})$ in an $n \times 64$ matrix V. The whole process of computing the n attention vectors, each of dimension 64, or equivalently the $n \times 64$ attention score matrix A, thus reduces to the two matrix multiplications

$$A = \left[\text{softmax}\left(\frac{QK^T}{\sqrt{64}}\right)\right] V, \tag{2.1}$$

where the softmax is applied row-wise to the $n \times n$ matrix $QK^T/\sqrt{64}$, and the result is multiplied with the $n \times 64$ matrix V; see Figure 2.1 (top right).

This mechanism of computing an attention score is carried out multiple times in parallel. For this reason, the architecture is called "multi-head attention". The rationale behind this is similar to using multiple filters in convolutional NNs: each of the attention head has its own trainable 64×512 matrices that it uses to find useful maps of the words to the query, key, and value vectors, thereby enabling each head to focus on learning different correlations. The paper uses eight attention heads, such that we get, for each input word, eight 64-dimensional attention score vectors. These are then simply concatenated to form a 512-dimensional vector.

Skip connection. Remember that the input word $|w\rangle_{EP}$ was represented by a 512-dimensional vector as well. This means that one can add the vector representing each input word to the output of the multi-head attention block, which realizes the aforementioned skip connection; the result is a vector that is a superposition of the input and the attention score.

These skip connections help avoid vanishing gradients. In addition to the skip connection, a layer norm is computed for these superposition vectors, which is another normalization step that helps avoid vanishing or exploding gradients; see Ref. [4].

Feed-forward NN. The final step in the encoder block is to feed these normalized 512-dimensional superposition vectors to a vanilla feed-forward NN that also has an output dimension of 512. This second map allows the encoder block to map the vectors to a "more useful" representation. The authors also allow bypassing this second map (i.e., they introduce another skip connection) and add the input to the feed-forward NN to the output, followed by another layer normalization step. Altogether, we can think of the encoder block as a complicated map $\mathbb{R}^{512} \to \mathbb{R}^{512}$ built from the concatenation $f_3(f_2(f_1(|w\rangle)))$, with

$$\begin{aligned} |a_i\rangle &:= f_1(|w_i\rangle_{EP}) = \left(\left[\text{softmax}\left(\frac{\mathcal{Q}\mathcal{K}^T}{\sqrt{64}}\right)\right]\mathcal{V}\right)_i, \\ |y_i\rangle &:= f_2(|a_i\rangle) = \text{LN}(|w_i\rangle + |a_i\rangle), \\ |r_i\rangle &:= f_3(|y_i\rangle) = \text{LN}(|y_i\rangle + |\text{NN}(|y_i\rangle)\rangle), \end{aligned} \tag{2.2}$$

where $\text{LN}(\cdot)$ denotes the layer norm and $\text{NN}(\cdot)$ denotes the map encoded by the feed-forward NN.

Each subsequent encoder block uses the output of the previous encoder block and repeats the above sequence of maps. The final output of the final encoder block will then be a sequence of n 512-dimensional vectors.

2.2.4 *The decoder*

The decoder works very similarly to the encoder in that it also uses multi-head attention blocks with skip connections and layer norms, plus a vanilla feed-forward NN at the end of each block. They are indicated by grey blocks in the decoder in Figure 2.1 (top left). The inputs to the decoder are its previous outputs (in the first step, a special "end-of-sentence" character is used as input). These inputs are processed as in the encoder case, i.e., each word is embedded into $\mathbb{R}^{512}$ using a $512 \times 300{,}000$ matrix, followed by adding a position encoding computed using sines and cosines. In the first decoder block, this embedded + position-encoded vector is then fed to the multi-head attention block, which computes the attention score as in (2.1), followed by a skip connection and layer norm. Instead of using the result directly as an input to a feed-forward NN as in the encoder case, the decoder has a second multi-head attention block. In this block, the output from the first attention block is used to compute the value. However, to compute the key and query vectors, one uses the output of the first encoder block instead. To prevent the network from "cheating" by looking ahead in the decoding phase, the scalar product between $\vec{q}_i$ and $\vec{k}_j$ masks out (sets to $-\infty$) all future positions. This ensures that the decoder can only attend to earlier positions in the output sequence. The output of the second multi-head attention block is then processed, as in the encoder case; cf. Eq. (2.2).

The above decoder block is also repeated multiple times. For the ith decoder block, the second multi-head attention block uses the output from the ith encoder block to compute the query and value vectors.

As a final step, the 512-dimension output from the last decoder block is multiplied by a $300{,}000 \times 512$ matrix (or whatever the size of the decoder dictionary is), and a softmax is applied. This gives a 300k-dimensional vector encoding a probability for each word in the dictionary. In a greedy decoding scenario, we simply choose the most likely word and add it to the output sentence. The output sentence plus this new word is then fed again to the decoder, where it gets embedded and positionally encoded, and so forth. In this way, one can produce long output sequences that can learn to pay attention to words that are quite far from the current word, which is sometimes necessary in natural language processing.

2.2.5 *Illustration of attention*

As an illustration, the authors of Ref. [2] look at the attention score of an attention head in the fifth encoder block for the input sentence: "The Law will never be perfect, but its application should be just — this is what we are missing, in my opinion"; see Figure 2.1 (bottom). One can see that many words pay strong attention to themselves — after all, they have a strong overlap with themsleves. But one can also see for example that the word "its" has a high attention score for the words "Law", meaning the attention mechanism is seemingly helping in anaphora resolution across six words.

References

[1] S. Bubeck, V. Chandrasekaran, R. Eldan, J. Gehrke, E. Horvitz, E. Kamar, P. Lee, Y. T. Lee, Y. Li, S. Lundberg, H. Nori, H. Palangi, M. T. Ribeiro, and Y. Zhang, "Sparks of artificial general intelligence: Early experiments with GPT-4," 2023.

[2] A. Vaswani, N. Shazeer, N. Parmar, J. Uszkoreit, L. Jones, A. N. Gomez, L. U. Kaiser, and I. Polosukhin, "Attention is all you need," in I. Guyon, U. V. Luxburg, S. Bengio, H. Wallach, R. Fergus, S. Vishwanathan, and R. Garnett (eds.), *Advances in Neural Information Processing Systems,* Vol. 30. Curran Associates, Inc., 2017.
[3] T. Mikolov, W.-T. Yih, and G. Zweig, "Linguistic regularities in continuous space word representations," in *Proceedings of the 2013 Conference of the North American Chapter of the Association for Computational Linguistics: Human Language Technologies,* pp. 746–751. Association for Computational Linguistics Atlanta, Georgia June, 2013.
[4] J. L. Ba, J. R. Kiros, and G. E. Hinton, "Layer normalization," 2016.

PART 2

METAHEURISTIC SEARCH TECHNIQUES: GENETIC ALGORITHMS AND QUANTUM ANNEALING

This part of the course introduces heuristic search methods — optimization techniques typically inspired by naturally occurring phenomena. We will focus on two such methods: genetic algorithms and quantum annealing. These represent opposite ends of the heuristic spectrum: in one, the problem of interest is largely decoupled from the mechanics of the search, while in the other, the problem is directly embedded within it. Although many other heuristic techniques exist, these two examples illustrate key advantages (e.g., broad applicability, potential speed-ups in suitable settings) and drawbacks (e.g., limited explainability) shared by many heuristic methods. In the tutorials, we will consider several discrete problems, such as Taxicab numbers and the Knapsack problem. The former is a well-known curiosity in number theory, while the latter is a classic problem in combinatorial optimization. Both are also morally analogous to certain discrete structures that arise in physics, such as anomaly cancellation conditions and partition function computations.

CHAPTER 3

GENETIC ALGORITHMS

STEVE ABEL AND THOMAS HARVEY

3.1 Introduction

The emergence of extremely large search spaces — not only in string theory but also in machine learning and condensed matter physics — has stimulated significant interest in heuristic search methods based on processes observed in nature. Metaheuristic techniques are among the most contentious approaches in artificial intelligence (AI). In general, they operate by mimicking natural systems that appear to produce remarkably effective solutions. They are controversial because, by their very nature, it is difficult to prove that they should work; yet experience repeatedly shows that they often do.

Nature, in fact, routinely copes with search spaces far larger than those encountered in any string-theoretic context. For example, the possible configurations of DNA are unimaginably vast, and yet biological evolution finds viable solutions within this "DNA search space" with apparent ease. We also observe that evolution must be highly efficient: empty ecological niches are filled rapidly, and countless instances of convergent evolution arise, such as echolocation in bats and dolphins or powered flight in insects and birds. The striking resemblance between the hummingbird hawk-moth and its avian namesake suggests that there may be a single optimal solution to the problem of extracting nectar from a flower while hovering in flight.

A final encouraging aspect of evolution, from which heuristic methods draw inspiration, is its "problem-agnostic" nature: regardless of the specific details or parameters of the challenge, evolution invariably finds a way.

In this chapter, we discuss genetic algorithms (GAs). These are methods for optimizing systems that can be traced back to Turing's ideas on machine intelligence in the 1950s [1]. GAs were the pioneering evolutionary approach, and nowadays there are many related techniques, for example evolutionary programming, differential evolution, and genetic programming (GP), as well as many cousins of these methods that have been inspired by other natural processes. In a GA, the purpose is to mimic natural selection by inserting a population of "creatures" into the search space, each of whose "genotypes" generate a different parameter choice. Each individual in the population represents an attempt at a solution which can be judged for its fitness. The creatures are then selected for breeding based on their fitness, and this over generations leads to a facsimile of evolution, with the aim that the population evolves to the desired solution.

There has been growing interest in this technique within the astrophysics, particle and string theory community, for example in Refs. [2–25], due to its often surprising efficacy. Indeed, it has been shown to be competitive with other methods such as reinforcement learning [26–28], and interestingly appears to operate in a somewhat complementary manner. In this lecture, we outline the procedure and also touch on its relation to GP.

3.2 Lecture

3.2.1 *The basic structure of genetic algorithms*

The GA approach is, as we have said, obviously inspired by evolution, which can for the purposes of our discussion be thought of as a method of solving the optimization problems that appear in nature. The modern GA method, containing all the crucial elements we would recognize today, was first developed by John Holland and de Jong building on Turing's ideas [29–38]. Useful reviews can also be found in Refs. [39–42]. They focused on the elements in evolution that they thought most vital. The first and most obvious element is the evolutionary pressure that allows good solutions to thrive while poor solutions flounder. The crucial aspect of this evolutionary pressure is that creatures in a population that are well suited to an environment have the luxury of breeding and passing on their genetic information. In rough terms, the fitter an individual, the more often it gets to breed. This is how information about the good fitness of such individuals gets written into the genetic code of the population and propagated through it over generations. Thus, according to Holland and de Jong, evolution should be mimicked by some kind of similar Markovian process.

The second obvious aspect of the evolutionary process is the fact that the offspring are not simply identical copies of the parents but are instead the result of some form of splicing that occurs during breeding. This gives the population as a whole the flexibility to rapidly adapt to environmental changes. If for example two parents carry a different and relatively rare beneficial gene, then one of their offspring may be lucky enough to carry both. That individual would be initially very rare in the population and therefore highly favoured for breeding (over, for example, its run-of-the-mill siblings). Thus, in combination with "selection", its genetic information and hence the new genetic configuration would very rapidly spread through the population.

The third element identified by Holland and de Jong is that of mutation. If, for example, our new super-fit individuals can breed *too* successfully, then there is a danger that the population as a whole stagnates. Every individual in the population will have acquired the same genetic material. Such a population would be unable to explore easily new configurations and would be poorly prepared for environmental shocks or for generating even fitter individuals. In order to avoid such stagnation, there must be some element of random variation introduced into the genetic information in the population.

In an astonishing piece of foresight, Holland and de Jong proposed that a system that featured these three elements of "selection", "crossover", and "mutation" would replicate the successes of evolution regardless of the details of the environment. In other words, one can encode virtually any optimization problem into an "environment" and devise an evolutionary framework with the above elements in order to try and solve it.

In this lecture, I describe a simple GA framework and then go on to describe some of the improvements that one can contemplate. Before I begin, I should remark that there are many GA packages, that are available for virtually every computer language. However, I do strongly recommend writing your own programs at least at the beginning because a simple implementation can be achieved with remarkably little code — maybe a couple of hundred lines suffices.

In order to consider the implementation of a GA, it is useful to have a toy optimization problem in mind. For pedagogical purposes, I usually like to focus at least initially on the problem of finding the maximum of a very wobbly two-dimensional function, such as

$$f(x,y) = 12\left(\cos\frac{3y}{2}\sin\frac{3x}{2} + x + y\right) - (x^2 + y^2),$$

without using calculus. We will see that there are various reasons that this problem is instructive.

How might one go about encoding this as a GA problem? First, we need a way to represent the problem as a genetic code. Typically, one uses binary genotypes, but the most straightforward way to do this for the current problem is to take the genotype of each "individual" in our population to be their x, y values. Thus, for pedagogical purposes, let us allow the individual alleles to take the values $0 \ldots 9$ and simply adopt the digits of the x, y values to be our alleles. In other words, the kth individual in the population with position $x = a.bcd\ldots$ and $y = g.hij\ldots$ would have a genotype that looks like

$$G_k = [abcd\ldots ghij\ldots],$$

where we take the genome to have length $2L$ so that our answer will be to L significant figures in each coordinate. The function itself is shown in Figure 3.1, together with a very artistically depicted population of individuals with randomly chosen $G_{k=1\ldots N_{\text{pop}}}$. Secretly, we know that the function has a maximum at $x = 7.23832\ldots$ and $y = 6.26348\ldots$, but if we pretend that we do not know this for the moment, then what we are ultimately hoping for is that our GA will end up producing an individual with genome

$$G_k = [723822\ldots 626348\ldots]. \tag{3.1}$$

I should stress that for pedagogical purposes only, the genotype in this example directly represents coordinates, but in realistic problems, it could encode any input parameters that one might desire.

The way that we are going to do this optimization is to work only with the phenotypes, Φ_k. The phenotypes are the physical representation of the genotypes, which in this particular case is simply the value of the function derived from G_k:

$$\Phi_k(G_k) = f\left(\sum_{\ell=0}^{L-1} G_k^\ell \times 10^{-\ell}, \sum_{\ell=L}^{2L-1} G_k^\ell \times 10^{-\ell}\right).$$

We will in addition need a way to judge the fitness of the individuals based on their phenotype in order to select them for breeding. In this example, we can take it to be the height itself, thus $F_k = \Phi_k$.

Even before we perform any evolution, we can make some general observations. The first is that even though the problem we are discussing in this pedagogical case is continuous, it has to be made discrete for the GA to operate. Conversely, there is no distinction in a GA between discrete and continuous problems. Thus, while it is commonplace to imagine a figure like that in Figure 3.1 and talk about a "fitness landscape" — and indeed, I myself will use this terminology occasionally — one should remember that this landscape is a discrete one. A second observation we can make is that there is a clear separation between the definition of the genotypes and the evaluation of the fitness of an individual. In order to evaluate the latter, one is instructed to take the genotype and perform several operations on it in order to turn it into parameters that can be fed into a function to derive

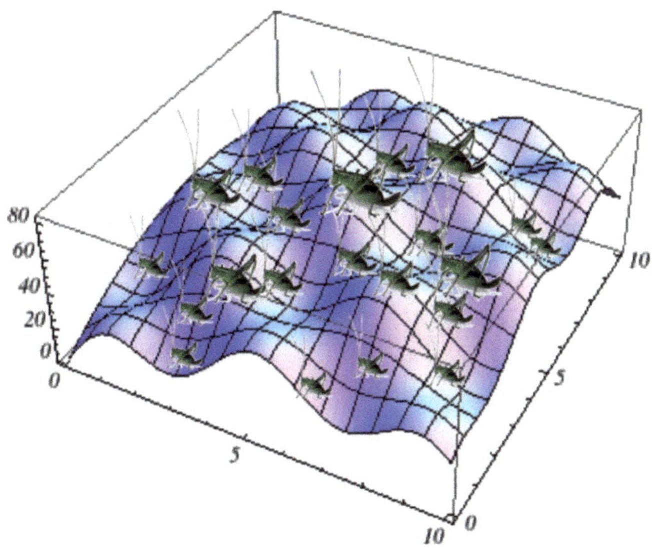

Fig. 3.1. Somewhat artistic depiction of a population being initialized on our toy function in order to find the maximum.

the phenotype and hence the fitness. The operation of extracting the phenotype is inevitably entirely decoupled from any operations that might be performed on the genotypes themselves. This is one of the beauties of the GA method: as far as the GA is concerned, the optimization problem that we are interested in solving is essentially a black box into which we can feed parameters derived from the genotype G_k.

Having created an initial randomly chosen population of N_{pop}, the subsequent flow of the GA can be visualized by means of the flowchart depicted in Figure 3.2 (taken from Ref. [43]).

The first step in the process is selection, in which individuals are first ranked based on their fitness values. In our toy example, we have assigned the fitness to be the function values $F_k = \Phi_k$ themselves. Thus, we begin at Genotype$'$ in Figure 3.2 by evaluating all the function values (also known as the phenotypes). For this simple example, equating this with the fitness effectively constitutes the "environment", but obviously in complicated problems, this step can be a multifaceted computation deriving from the

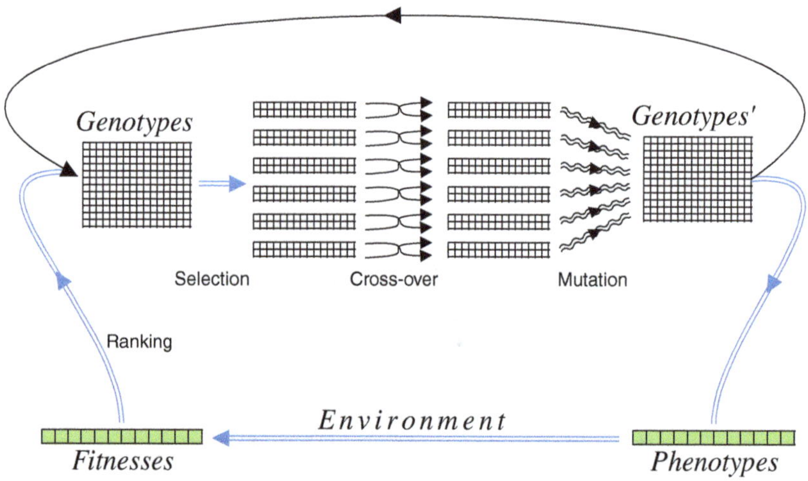

Fig. 3.2. The classical GA is a directed graph with two loops. The initial population of N_{pop} genotypes (on the right) give rise to phenotypes, which are in turn used to evaluate the fitnesses of the N_{pop} individuals in the "environment", which are then collected as a ranked population of parent genotypes (on the left). The fitness ranking of an individual determines the probability of it being selected to take part in a breeding pair. $N_{\mathrm{pop}}/2$ breeding pairs are formed in this manner, and then crossover and mutation yield an entirely new generation, and the process is repeated.

phenotype of each individual. The population can then be ranked according to the fitness values. Following ranking, individuals are chosen for breeding with probabilities that are often for convenience taken to vary linearly with their ranks. The probability (P_k) of selecting the kth individual for breeding is given by

$$P_k = \frac{2}{(1+\alpha)N_{\text{pop}}}\left(1 + \frac{N_{\text{pop}} - k}{N_{\text{pop}} - 1}(\alpha - 1)\right).$$

Notably, P_1, representing the top individual, is α times greater than $P_{N_{\text{pop}}}$, which corresponds to the least fit individual. Typically, α (which can be thought of as the equivalent of the "learning rate") falls in the range $2 \leq \alpha \leq 5$. This strategy ensures that both the fittest and less fit individuals have opportunities to contribute to the next generation, maintaining diversity in the gene pool.

Breeding is realized through an M-point crossover, where two binary sequences are cut at M random points, and the cut sections are swapped in an alternating fashion. Typically, a single-point crossover suffices in fact. Finally, a mutation stage is implemented to prevent population stagnation, with a small portion of digits being randomly selected and flipped to another value. A common mutation rate is 1%, but the optimal value is highly problem dependent.

This process generates a new set of genotypes: we are now back at Genotypes' in the figure, with an entirely new population having replaced the old one. Then, the whole process is iterated, with phenotype production, fitness calculation, ranking, selection, breeding, and mutation repeated over many generations. Figure 3.2 shows this process as a graph with two loops. This is to emphasize the procedure's most important feature, namely that within the lower loop one is, as promised, working entirely with the phenotypes. This part of the graph is where the problem to be solved is "bolted on" to the genetic algorithm. Thus, one can as mentioned simply import previously developed "black-box" routines that take the set of parameters generated by the genotype as input. This modular structure makes the framework exceedingly easy to adapt.

To summarize the algorithm, the core elements that drive the evolution of the population in GAs are threefold: selection, breeding, and mutation. Written out as pseudocode, the procedure is as shown in Algorithm 1 (where we also include *elitism*, namely the preservation of the previous fittest individual):

Algorithm 1: Typical GA

```
for population do
    Initialize genotype
end for
while generations < gen − number do
    for population do
        Find phenotype
        Rank individuals by fitness
        Elitism: Copy fittest individual
    end for
    for new population do
        Using fitness select breeding pair
        Crossover genotypes to create new individuals
        Mutate new individuals
    end for
    Elitism: Paste previous fittest over bottom-ranked
end while
```

The end result of such an implementation (modulo the practical modifications for more realistic problems that I am about to mention) is that over multiple generations of evolution, often several hundred, the GA yields a final population in which a significant portion of individuals represents viable solutions to the problem at hand. For our toy example, the population ends up clustering as in Figure 3.3.

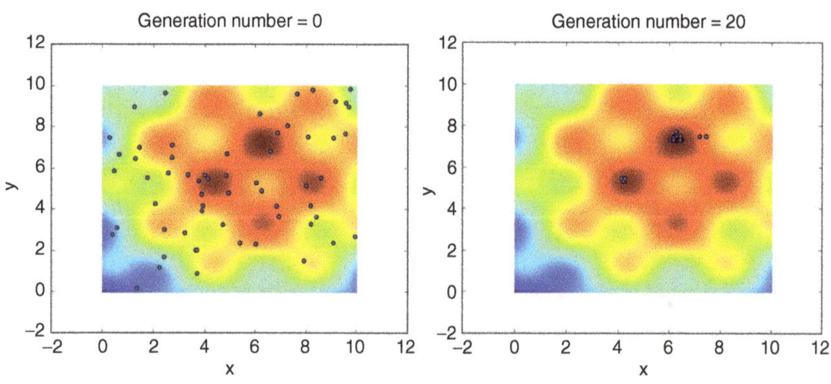

Fig. 3.3. Evolution of a population of 60 individuals in our toy problem.

3.2.2 *Practicalities, improvements, and optimization*

Of course, not all problems are as simple as our pedagogical toy problem, and to tackle most realistic problems, the configuration of a GA has to be significantly more finessed. Thus, let us now discuss the practicalities and modifications (most of which do not significantly change the philosophy of the evolutionary approach). First, although for pedagogical purposes we used a decimal encoding, a binary encoding produces better results. Indeed, with a binary encoding, the optimum size of the population N_{pop} depends only logarithmically on the genotype length ℓ, with a formal lower bound of [36]

$$N_{\text{pop}} \gtrsim 1 + \frac{\log(\ell/\log P_*)}{\log 2},$$

which ensures that both possible values for each allele should appear in the initially randomized population with some representative probability P_*. In addition, the mutation rate can be optimized — although there do exist expressions for the optimum values of both these parameters [36], typically the optimum values depend on the precise problem at hand and would be finessed numerically. This optimization of the meta-parameters can be automated using for example `Optuna` [44] (see also Ref. [45]).

The fitness function need not always be based directly on ranking. However, ranking has some advantages. For example, it is robust against almost flat distributions in which the fitnesses tend to be very close together. Often, *crowding penalties* are implemented in the fitness function to attempt to prevent the population stagnating. In addition to *roulette-wheel* selection, there are many alternatives, including *steady-state, stochastic, tournament,* and *Boltzmann selection* (which has some moral overlaps with simulated annealing). In addition to these modifications, it is often advantageous to implement the aforementioned *elitism*, in which the fittest individual is maintained in the population and not mutated. In addition, when there are multiple criteria that are being optimized, the algorithm can be made more efficient by using Pareto optimization (again, see Ref. [45]).

Finally, modifications can be made to the phenotype as well (going partly against the philosophy of GAs). One example that was introduced in Ref. [6] is the so-called *creep-mutation* which operates partially also on the phenotype in order to overcome the so-called Hamming walls (which can arise when very close phenotypes are very far away

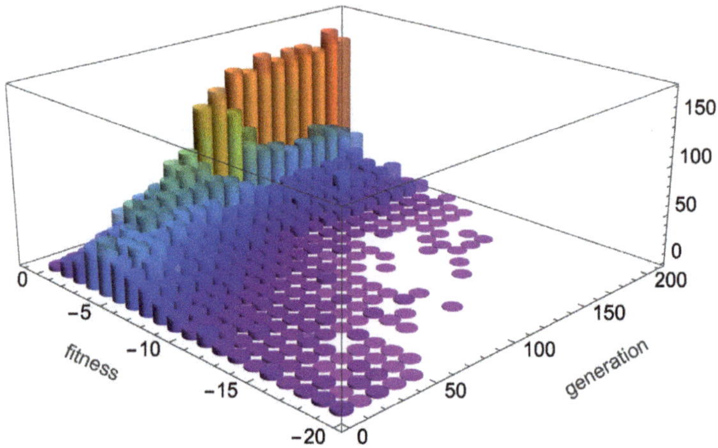

Fig. 3.4. Evolution of a population of string models.
Source: Figure taken from Ref. [26].

in "Hamming space", in the sense that the very close binary numbers $100\cdots00$ and $011\cdots11$ have maximally different digits).

In Figure 3.4, we show a more realistic example taken from Ref. [26] that utilizes some of these features (in particular, binary encoding and elitism). The particular problem being considered here is to find favourable Calabi–Yau string models in a particular (bicubic monad bundle) configuration. The problem boils down to finding solutions to a set of Diophantine conditions. The figure shows the typical GA behaviour: the average fitness within the population grows very quickly even while it retains a good spread of individuals. In this state, the population begins to find solutions exceedingly quickly, indeed requiring much shorter training times than, for example, reinforcement learning [26]. In the example considered here, the search space is about 10^{12}, but the GA is capable of finding 48 solutions after only 50,000 attempts, while a random search over millions of states would typically lead to no solutions at all. The efficacy can indeed be astonishing: similar studies show the ability to find most viable models in a search space of 10^{29} after visiting only a tiny fraction (10^{-19}) of the search space [46].

3.2.3 *Why does it work?*

The reason for the efficacy of GAs has been controversial. Nowadays, we can recognize an interesting mix of elements that can be found in other optimization methods. The fitness function for example is simply a kind of cost function, and the parameter α clearly is in some sense the "learning rate" of the algorithm. Pursuing this direction of thought, the idea of fitness–distance correlation, which was explored by Jones, Forrest *et al.* [33–35,39,47], suggests some parallels with the gradient-descent methods that are typically employed to train neural networks. However, one can also see that GAs must confer their own additional advantages beyond simple fitness–distance gradients: this is because very strong correlation between fitness (which is a phenotype property) and distance in the search space would make the genotype representation superfluous because the dynamics of the flow could simply be mapped directly to the phenotype. This is clearly then not the whole story for GAs since, in addition, small changes in the genotype can result in huge changes in the phenotype (the converse advantage to the aforementioned disadvantage of Hamming walls). In this sense, the GA can more readily sample the whole space before it begins to converge.

Holland's original schema theory idealized this behaviour in terms of hidden building blocks within the genotype. A schema is a representation of some crucial set of digits that confers some favourable characteristic. For our toy example, it might look something like $S = 72 * * * \cdots * 62 * * * \cdots$. This example has four entries that we deem to be important because they determine the leading two digits of the x, y coordinates. Hence, we say that the schema is order 4. The entries that we care a lot less about are labelled with a wildcard $*$. Clearly, this schema will be far more important in determining the phenotype than the others. Although for most problems the important schemata will not be as obvious as they are in this one, Holland argued that they are operating behind the scenes anyway, even if they are not readily identifiable. The probabilistic arguments of Holland (summarized in Ref. [13]) then suggest that schemata of this kind are important because selection favours the propagation of shorter strings of data: small subsections of the genome that confer fitness dominate first, and once they are shared by the majority of the population, crossover does not affect them. As a result, so the theory goes, if the schema S confers fitness on an individual, then it will grow exponentially within the population until an equilibrium is reached, where the attraction to the solution balances against the repulsion due to the mutation. In this simple example, we can indeed observe individuals beginning to "line up" along the smaller second-order schemata: $S = 72 * * * \ldots$ and $S = \ldots * 62 * * * \ldots$

The probabilistic schema argument suggests further that a GA incorporates and balances competing forces. Selection and breeding tend to produce convergence around local maxima in the fitness landscape, drawing the population in over generations. On the other hand, the effect of mutation is to push the population away from local maxima (on average), so that as a whole it can explore the entire parameter space. The power of GAs is then that they are sensitive to the entire landscape but can simultaneously respond to and converge on interesting regions.

The schema argument fell temporarily out of favour in the 1990s: see Ref. [42] for discussions of why, but essentially it was because it does not consider the constructive effects of crossover and it assumes a constant contribution to the fitness of a given schema. By contrast, the population is not static, so that changes in the rest of the genotype tend to alter the relative fitness of the schema after the first few generations in a run. Despite such objections, the schema idea was reconsidered by Poli *et al* in a suite of papers (see Ref. [42] for a comprehensive review), which developed the so-called *exact* schema-based models and showed their relation to Markov chain processes. There have been several other approaches for understanding GAs, for example statistical mechanical approaches [48] or stochastic differential equations [49]. In the former approach, one assigns the fitness function with a Boltzmann weighting, while the latter approach is restricted to simple cases such as the so-called onemax problem, which is equivalent to solving the one-dimensional Ising model. In these cases, it is likely that the setup of the configuration influences the results, so that the conclusions are not general but really only applicable to the very specific configurations used to frame the argument. Despite all its limitations then, the schema theory remains as possibly the most interesting interpretation of the behaviour of the GA.

3.2.4 *Towards explainability: Genetic programming*

Finally, it is worth mentioning a related evolutionary algorithm, namely GP. One of the drawbacks of a GA (and indeed many other techniques) is that it is difficult to understand the results. If one seeks only a solution to a given problem, then this is of course of no consequence. However, for many problems, one would like to implement a form of symbolic regression (for example, one might wish to train on data to find an analytic expression for a classification curve). The GP method for such symbolic regression was instigated by Koza in Ref. [50]. A GP operates with trees of

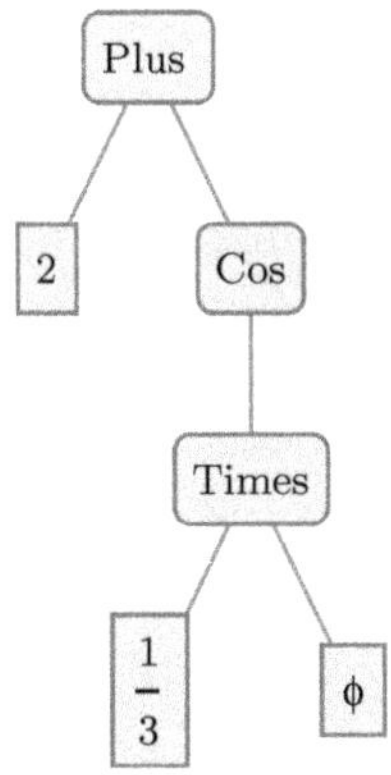

Fig. 3.5. Tree form of the function $f(\phi) = 2 + \cos(\phi/3)$.

symbolic expressions and numerical factors. For example, the tree form of the simple cosine function $f(\phi) = 2 + \cos(\phi/3)$ is shown in Figure 3.5.

The algorithm for a GP follows that of the GA, operating on a population of such trees. The difference of course arises at the breeding stage, where in the GP the crossover is implemented by swapping subtrees. This operation can take many more forms than a simple one-point exchange of subtrees (the so-called homologous crossover), with subtrees possibly being moved to entirely different positions in the tree. The importance of mutation is less clear cut in GPs and was even omitted from Koza's original work [50]; however, typically, it is implemented as well. A comprehensive survey of GPs can be found in Ref. [51]. In particular, the benchmark for GPs at the time of writing seems to be the Operon package (and its Python wrapped version Pyoperon) [52]. It was shown that this method can successfully be applied to perform global fit analysis of beyond Standard Model physics in Ref. [53].

3.3 Conclusions

In summary then, it is clear that GAs (and, to a lesser extent, their cousins GPs) can be efficiently used to tackle many computationally difficult problems, and they have proven to be one of the most effective metaheuristic search methods. They simply excel at problems that are multidimensional or that have many local minima, thanks to their three crucial components: selection, breeding, and mutation. Their great advantage is that they tend to converge very rapidly at first and can produce solutions very rapidly, requiring less training time than many other methods before they start to produce solutions. However, the rate of convergence tends to slow down as they converge on solutions. Therefore, they can sometimes be most effective when used in conjunction with other heuristic methods, for example with the Nelder–Mead optimization. Finally, GP and symbolic regression can add some aspects of explainability to such evolutionary algorithms. In particular, they can provide fresh inspiration for model building and allow far more efficient fitting of proposed models. However, they remain far less explored.

3.4 Tutorial

3.4.1 *Example code*

Throughout this tutorial, we assume that you have access to a `Python` installation with both `NumPy` and `matplotlib` available.

We begin by setting up a GA to solve a simple problem involving Diophantine equations. In particular, we will use a GA to search for *taxicab numbers*, which are numbers that can be written as the sum of two positive integer cubes in at least two distinct ways. More precisely, if T is a taxicab number, then there exist $a, b, c, d \in \mathbb{Z}^+$ with $a \neq c, d$ such that

$$T = a^3 + b^3 = c^3 + d^3.$$

The exercises will involve modifying this code to solve several other problems using GAs. We begin by importing the necessary Python packages and defining a number of useful hyperparameters:

```
import numpy as np
import matplotlib.pyplot as plt
import random

numBits = 5*4
alpha = 5
popSize = 200
numGens = 200
mRate = 1/25
elite = True
```

We next define a few useful functions, including one that removes duplicate entries from a list and another that converts a binary string into an integer expressed in base 10:

```
def removeDuplicates(lst):
    unique_list = []
    for item in lst:
        if item not in unique_list:
            unique_list.append(item)
    return unique_list

def bitstointeger(binary):
    number = 0
    for b in range(len(binary)):
        number += binary[b] * 2**b
    return number
```

We now introduce a class that defines an individual (creature), together with an associated fitness function. These are given as follows:

```
#This fitness function is maximised (=0), for taxicab numbers
def fitFunc(bits):
    split = int(numBits/4)
    x = bitstointeger(bits[0:split])
    y = bitstointeger(bits[split:2*split])
    z = bitstointeger(bits[2*split:3*split])
    w = bitstointeger(bits[3*split:4*split])

    #Sum of two cubes in two ways
    fitness = -(x**3 + y**3 - z**3 - w**3)**2

    #Those two ways must be different (When the above
        condition is true, only need to check x==z and x==w)
    #Choosing 500 took some tuning
    if x == z:
        fitness -= 500
    if x == w:
        fitness -=500

    return fitness, [x,y,z,w]
```

```
class creature:
    #Class has three components:
        #bits: List of bits defining the creature (genotype)
        #vals: The four integers to be cubed (phenotype)
        #fitness: The fitness of the creature, calculated
        #    using the above fitness function

    def __init__(self, bits):
        if bits==[]:
            self.bits = []
            for i in range(numBits):
                self.bits.append(random.randint(0,1))
        else:
            self.bits = bits
        self.fitness, self.vals = fitFunc(self.bits)
```

To run the GA, we define functions for both crossover and mutation:

```
#For this example, we will only consider a single crossover
def crossover(ind1,ind2):
    #randomly choose the crossover point
    split = random.randint(0,numBits)
    #generate the two new creatures and return them
    newBits1 = ind1.bits[0:split] + ind2.bits[split:numBits]
    newBits2 = ind2.bits[0:split] + ind1.bits[split:numBits]
    return creature(newBits1),creature(newBits2)
```

```
def mutate(ind):
    bits = ind.bits
    #loop over each bit in the bit string
    for split in range(0,numBits):
        # Mutate bit if random draw is below mutation rate
        if random.random() < mRate:
            if bits[split]==0:
                bits[split]=1
            else :
                bits[split]=0
    #return the new creature
    return creature(bits)
```

We can test these functions by generating two random individuals and applying crossover and mutation:

```
state1 = creature([])
state2 = creature([])
print("Init:     ",state1.bits,state2.bits)
new1,new2 = crossover(state1,state2)
print("Cross:    ",new1.bits,new2.bits)
new3 = mutate(new1)
new4 = mutate(new2)
print("Mutate:   ",new3.bits,new4.bits)
```

We now have all the components needed to run the GA, which proceeds as follows:

```
# Generate the initial population and set the probability
# weights for selecting individuals for breeding
pop = [[]]
weights = []
for i in range(0,popSize):
    pop[0].append(creature([]))
    # Note the extra minus one from the formula due to
    #    zero-based indexing
    weights.append((2/(1+alpha))*
        (1+(popSize-i-1)*(alpha-1)/(popSize - 1))/popSize)

getFitness = lambda a : a.fitness

#Sort the initial population
pop[0] = sorted(pop[0],key=getFitness)[::-1]

#Loop over all populations
for r in range(0, numGens):
    newPop = []
    for i in range(0,int(popSize/2)):
        #randomly draw from the population using the weights
        k1 = np.random.choice(range(popSize), 1, p=weights)[0]
        k2 = np.random.choice(range(popSize), 1, p=weights)[0]
```

```
        #Crossover and mutate the two creatures to from two
            new ones
        new1, new2 = crossover(pop[r][k1],pop[r][k2])
        new1 = mutate(new1)
        new2 = mutate(new2)
        newPop.append(new1)
        newPop.append(new2)

    #If eliteism is enabled, add the fittest individual from
    #    the previous generation to the next
    if elite :
        newPop[-1] = pop[r][0]

    #Sort the new population by fitness and append to the list
    #    of populations
    newPop = sorted(newPop,key=getFitness)[::-1]
    pop.append(newPop)
```

If the GA has run successfully, there should be terminal states at the end of training. We now analyze the resulting population:

```
avFit = []
bestFit = []
terms = []
#Loop over entire population
for n in range(popSize):
    #Calculate the average fitness in this generation
    avFit.append(np.mean(list(map(lambda a : a.fitness,
        pop[n]))))
    #Save the highest fitness in this generation
    bestFit.append(pop[n][0].fitness)
    #Save the terminals in this generation
    terms = terms + [p.vals for p in pop[n] if p.fitness==0]
terms = removeDuplicates(terms)
```

We can plot the average fitness across generations:

```
plt.plot(avFit)
```

and similarly plot the best fitness:

```
plt.plot(bestFit)
```

We can also check the taxicab condition for the best solution by evaluating:

```
terms[0][0]**3+terms[0][1]**3-terms[0][2]**3-terms[0][3]**3
```

This should return zero.

3.4.2 *Problems*

Problem 1

(A) Modify the code above so that the GA searches for approximate minima of a given multivariable function, where the variables lie within a specified interval. The minimum may occur in the interior or on the boundary of the domain.
(B) Extend your solution so that the GA searches for both local minima and maxima of a given function. You will likely need several separate GA runs to find all of them.

Problem 2

(A) Modify the code to solve a Knapsack problem. In this problem, we are given a list of items with weights and values, together with a maximum weight capacity. The goal is to maximize the total value while respecting the weight constraint.
(B) Now, allow for multiple bags and ignore item values. Use a GA to minimize the total number of bags required to carry all items.
(C) Consider the possibility that a single bag may be lost. Rewrite the fitness function to minimize the value lost if any one bag goes missing while still using the minimum number of bags.

Problem 3

Write your own version of a GA in any programming language of your choice.

Problem 4

A magic square is a 3×3 grid of unique integers such that all rows, columns, and diagonals sum to the same value. Use a GA to search for a *magic square of squares*, that is, a magic square where every entry is a distinct perfect square.

This problem is for your own enjoyment, as no magic square of squares is known to exist, and there is evidence that none can exist.

References

[1] A. M. Turing, "I.—Computing machinery and intelligence," *Mind* **LIX**, 433, 1950.

[2] P. Charbonneau, "Genetic algorithms in astronomy and astrophysics," *ApJS* **101**, 309, 1995.

[3] A. Yamaguchi and H. Nakajima, "Landau gauge fixing supported by genetic algorithm," *Nucl. Phys. Proc. Suppl.* **83**, 840, 2000 [hep-lat/9909064].

[4] T. S. Metcalfe and P. Charbonneau, "Stellar structure modeling using a parallel genetic algorithm for objective global optimization," *J. Comput. Phys.* **185**, 176, 2003 [astro-ph/0208315].

[5] B. C. Allanach, D. Grellscheid, and F. Quevedo, "Genetic algorithms and experimental discrimination of SUSY models," *JHEP* **07**, 069, 2004 [hep-ph/0406277].

[6] Y. Akrami, P. Scott, J. Edsjo, J. Conrad, and L. Bergstrom, A profile likelihood analysis of the constrained MSSM with genetic algorithms, *JHEP* **04**, 057, 2010 [0910.3950].

[7] S. Nesseris and J. Garcia-Bellido, "A new perspective on dark energy modeling via genetic algorithms," *J. Cosmol. Astropart. Phys.* **2012**, 033, 2012.

[8] J. Blåbäck, U. Danielsson, and G. Dibitetto, "Fully stable dS vacua from generalised fluxes," *JHEP* **08**, 054, 2013 [1301.7073].

[9] C. Damien, L. R. Diaz-Barron, O. Loaiza-Brito, and M. Sabido, "Slow-roll inflation in non-geometric flux compactification," *J. High Energy Phys.* **2013**, 6, 1–29, 2013.

[10] C. Damian and O. Loaiza-Brito, "More stable de Sitter vacua from S-dual nongeometric fluxes," *Phys. Rev. D* **88**, 046008, 2013 [1304.0792].

[11] J. Blåbäck, U. Danielsson, and G. Dibitetto, "Accelerated universes from type IIA compactifications," *JCAP* **03**, 003, 2014 [1310.8300].

[12] J. Blåbäck, D. Roest, and I. Zavala, "De Sitter vacua from nonperturbative flux compactifications," *Phys. Rev. D* **90**, 024065, 2014 [1312.5328].

[13] S. Abel and J. Rizos, "Genetic algorithms and the search for viable string vacua," *JHEP* **08**, 010, 2014 [1404.7359].

[14] R. Hogan, M. Fairbairn, and N. Seeburn, "GAz: A genetic algorithm for photometric redshift estimation," *Mon. Not. R. Astron. Soc.* **449**, 2040, 2015 [1412.5997].

[15] J. K. Behr, D. Bortoletto, J. A. Frost, N. P. Hartland, C. Issever, and J. Rojo, "Boosting Higgs pair production in the $b\bar{b}b\bar{b}$ final state with multivariate techniques," *Eur. Phys. J. C* **76**, 386, 2016 [1512.08928].

[16] F. Ruehle, "Evolving neural networks with genetic algorithms to study the string landscape, *JHEP* **08**, 038, 2017 [1706.07024].

[17] S. Abel, D. G. Cerdeño, and S. Robles, "The power of genetic algorithms: What remains of the pMSSM?" [1805.03615].

[18] K. Bull, Y.-H. He, V. Jejjala, and C. Mishra, "Machine learning CICY threefolds," *Phys. Lett. B* **785**, 65, 2018 [1806.03121].

[19] A. Cole, A. Schachner, and G. Shiu, "Searching the landscape of flux vacua with genetic algorithms," *JHEP* **11**, 045, 2019 [1907.10072].

[20] A. E. Cole, "Identifying and exploiting structure in cosmological and string theoretic data," Ph.D. Thesis, University of Wisconsin, Madison, 2020.

[21] S. AbdusSalam, S. Abel, M. Cicoli, F. Quevedo, and P. Shukla, "A systematic approach to Kähler moduli stabilisation," *JHEP* **08**, 047, 2020 [2005.11329].

[22] N. Cabo Bizet, C. Damian, O. Loaiza-Brito, D. K. M. Peña, and J. A. Montañez Barrera, "Testing swampland conjectures with machine learning," *Eur. Phys. J. C* **80**, 766, 2020 [2006.07290].

[23] I. Bena, J. Blåbäck, M. Graña, and S. Lüst, "Algorithmically solving the tadpole problem," [2103.03250].

[24] G. J. Loges and G. Shiu, "Breeding realistic D-brane models," *Fortsch. Phys.* **70**, 2200038, 2022 [2112.08391].

[25] P. Berglund, Y.-H. He, E. Heyes, E. Hirst, V. Jejjala, and A. Lukas, "New Calabi-Yau manifolds from genetic algorithms," [2306.06159].

[26] S. Abel, A. Constantin, T. R. Harvey and A. Lukas, "Evolving heterotic gauge backgrounds: Genetic algorithms versus reinforcement learning," *Fortsch. Phys.* **70**, 2200034, 2022 [2110.14029].

[27] A. Cole, S. Krippendorf, A. Schachner, and G. Shiu, "Probing the structure of string theory vacua with genetic algorithms and reinforcement learning," in *35th Conference on Neural Information Processing Systems*, 11, 2021 [2111.11466].

[28] S. A. Abel, A. Constantin, T. R. Harvey, and A. Lukas, "Cosmic inflation and genetic algorithms," *Fortsch. Phys.* **71**, 2200161, 2023 [2208.13804].

[29] J. Holland, *Adaptation in Natural and Artificial Systems*. The MIT Press reprinted, Cambridge, MA, USA, 1992.

[30] K. D. Jong, "An analysis of the behavior of a class of genetic adaptive systems," Ph.D. Thesis, University of Michigan, Ann Arbor, MI, USA, 1975.

[31] D. E. Goldberg, *Genetic Algorithms in Search, Optimization and Machine Learning*. Addison-Wesley, Boston, MA, USA, 1989.

[32] J. Holland, "The royal road for genetic algorithms: Fitness landscapes and GA performance," in F. J. Varela and P. Bourgine (eds.), *Toward a Practice of Autonomous Systems: Proceedings of the First European Conference on Artificial Life*. MIT Press, 1992.

[33] S. Forrest and M. Mitchell, "What makes a problem hard for a genetic algorithm? Some anomalous results and their explanation," *Mach. Learn.* **13**, 285, 1993.

[34] T. Jones and S. Forrest, "Fitness distance correlation as a measure of problem difficulty for genetic algorithms," in L. J. Eshelman (ed.), *Proceedings of the 6th International Conference on Genetic Algorithms*, 1995.

[35] P. Collard, A. Gaspar, M. Clergue, and C. Escazu, "Fitness distance correlation, as statistical measure of genetic algorithm difficulty, revisited," 1998.

[36] C. Reeves and J. Rowe, *Genetic Algorithms: Principles and Perspectives*. Springer, Princeton, New Jersey, USA, 2002.

[37] R. Haupt and S. Haupt, *Practical Genetic Algorithms*, 2nd edn. Wiley, Hoboken, New Jersey, USA, 2004.

[38] Z. Michalewicz and D. B. Fogel, *How to Solve It: Modern Heuristics*, 2nd edn. Springer, Princeton, New Jersey, USA, 2004.

[39] T. Jones, "Evolutionary algorithms, fitness landscapes and search," PhD Thesis.

[40] P. Charbonneau, "An introduction to genetic algorithms for numerical optimization," Technical Report. TN-450+IA, National Center for Atmospheric Research, March 2002.

[41] J. McCall, "Genetic algorithms for modelling and optimisation," *J. Comput. Appl. Math.* **184**, 205, 2005.

[42] D. White, "An overview of schema theory."

[43] S. Abel, L. A. Nutricati, and M. Spannowsky, "A genetic quantum annealing algorithm," [2209.07455].

[44] T. Akiba, S. Sano, T. Yanase, T. Ohta, and M. Koyama, "Optuna: A next-generation hyperparameter optimization framework," in *Proceedings of the 25th ACM SIGKDD International Conference on Knowledge Discovery and Data Mining*, 2019.

[45] F. A. de Souza, M. Crispim Romão, N. F. Castro, M. Nikjoo, and W. Porod, "Exploring parameter spaces with artificial intelligence and machine learning black-box optimization algorithms," *Phys. Rev. D* **107**, 035004, 2023 [2206.09223].

[46] S. A. Abel, A. Constantin, T. R. Harvey, A. Lukas, and L. A. Nutricati, "Decoding nature with nature's tools: Heterotic line bundle models of particle physics with genetic algorithms and quantum annealing," [2306.03147].

[47] T. Jones and S. Forrest, "Fitness distance correlation as a measure of problem difficulty for genetic algorithms," Working Papers 95-02-022, Santa Fe Institute, February 1995.

[48] A. Prügel-Bennett and J. L. Shapiro, "Analysis of genetic algorithms using statistical mechanics," *Phys. Rev. Lett.* **72**, 1305, 1994.

[49] J. Heredia, "Modelling evolutionary algorithms with stochastic differential equations," *Evol. Comput.* **26**, 1, 2017.

[50] J. R. Koza, *Genetic Programming: On the Programming of Computers by Means of Natural Selection*. MIT Press, Cambridge, MA, USA, 1992.

[51] F. O. de Franca, M. Virgolin, M. Kommenda, M. S. Majumder, M. Cranmer, G. Espada *et al.*, "Interpretable symbolic regression for data science: Analysis of the 2022 competition," 2023.

[52] B. Burlacu, G. Kronberger, and M. Kommenda, "Operon c++: An efficient genetic programming framework for symbolic regression," in R. Allmendinger *et al.* (eds.), *Proceedings of the 2020 Genetic and Evolutionary Computation Conference Companion*, GECCO'20, pp. 1562–1570. Association for Computing Machinery, 8–12 July 2020, DOI: 10.1145/3377929.3398099.

[53] S. AbdusSalam, S. Abel, and M. Crispim Romão, "Symbolic regression for beyond the standard model physics," arXiv:2405.18471 [hep-ph].

CHAPTER 4

QUANTUM ANNEALING

STEVE ABEL AND LUCA A. NUTRICATI

4.1 Lecture

We now turn to our second kind of metaheuristic approach, namely annealing. Specifically, in this course, we shall discuss quantum annealing.

A quantum annealer is a kind of quantum computer that, even though non-universal, has been designed specifically for the task of minimizing a potential [1–12]. Our focus is on quantum annealers, but we should remark that much of our discussion can also be applied to other quantum computing paradigms, such as gate quantum computers. (Indeed, much of the discussion will also apply to classical thermal annealing.) However, at the time of writing, accessible quantum annealers, for example the devices provided by D-Wave Systems [13], have the significant advantage over quantum gate computers that they offer systems with several thousand qubits [14].

The general principle behind a quantum annealer is to solve problems by reformulating their solution as the minimization of a function H of logical spin variables which can take values $\sigma_\ell = \pm 1$, where ℓ labels the sites. The function, H, is called the *problem-Hamiltonian*. It is quadratic in the spin variables and takes the following form:

$$H(\sigma) = \sum_\ell h_\ell \sigma_\ell + \sum_{\ell m} J_{\ell m} \sigma_\ell \sigma_m. \tag{4.1}$$

Ultimately, in a quantum annealer, the logical spin variables here will correspond to physical *qubits*. Meanwhile, the matrix $J_{\ell m}$ and vector h_ℓ

correspond to couplings and biases, respectively, which can be set by hand. (Indeed, the quadratic form of H is somewhat dictated by the physical architecture of the annealer due to the fact that the qubits live on superconducting loops, which can only couple where they overlap.) It is through these couplings that the particular problem can be encoded as an optimization problem of H, such that the absolute minimum of H is a solution to the problem. We can postpone further discussion of how one performs the minimization of $H(\sigma)$ in practice and, for the moment, concentrate on how one can use $H(\sigma)$ to encode a problem. This part of the lecture is obviously relevant to any annealer based on such a generalized Ising model.

As a simple example of encoding a problem in $H(\sigma)$, consider the small network shown in Figure 4.1. The problem to be solved is the following: what is the maximum number of nodes that can be coloured on the given network if we are not allowed to have two coloured nodes linked together? A couple of example attempts on a sample network are shown in Figure 4.1 (and actually, the figure on the left is the solution for that particular network).

This is an NP-hard problem (see, for example, Ref. [15]; for recent reviews related to this context, see Refs. [16, 17]) that has an obvious exponential-time algorithm — namely, exhaustively scanning all possible values of σ_ℓ — which grows as 2^N, where N is the number of sites on the network.

In order to embed this problem into $H(\sigma)$, our first task is to decide how to assign colours to σ_ℓ values. Let us identify $\sigma_\ell = +1$ with coloured nodes and $\sigma_\ell = -1$ with uncoloured nodes. Then, we must insert the objective, namely to colour in as many nodes as possible, into $H(\sigma)$, which can be done by assigning a large negative bias to the h-coupling for every site. As we do not have a preference for which nodes are coloured, these biasing couplings are taken to be degenerate. Next, we also have to add a penalty

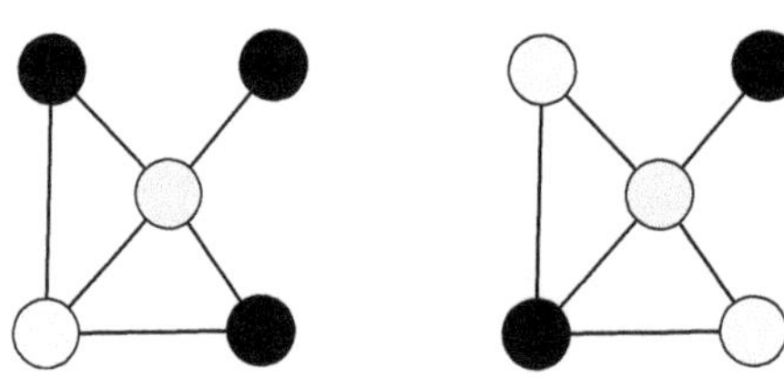

Fig. 4.1. Colouring a network with no two coloured nodes linked.

to H that prevents the network from colouring adjacent nodes. As this is a constraint, it should dominate over the biasing. That is, the penalty should outweigh the bias in deciding if a node is to be coloured or not. This can be achieved by adding a term proportional to the product $(1+\sigma_\ell)(1+\sigma_m)$ into H for every pair of linked sites. This product is zero if either of the sites is uncoloured (i.e., has $\sigma = -1$) but gives a penalty if both sites are coloured as required. Thus, the problem-Hamiltonian can be taken to be

$$H(\sigma_\ell) = -\Lambda \sum_\ell \sigma_\ell + \Lambda' \sum_{\text{linked pairs } \{\ell,m\}} (1+\sigma_\ell)(1+\sigma_m),$$

where Λ and Λ' are degenerate couplings that have $\Lambda < 4\Lambda'$.

Network problems can very easily be encoded in this manner. As another example problem, suppose there are N^2 students and that $N^2/2$ of them are ill: the students are to sit at N^2 desks arranged in a square grid, and we would like to seat them so that as few ill students as possible are sitting adjacent to healthy students as possible. In a similar fashion to the previous example, we first assign $\sigma = +1$ to stand for a desk at which an ill person should sit, while $\sigma = -1$ stands for a desk at which a healthy student sits. As the problem is arranged as an $N \times N$ grid, we will define indices as $\ell N + i$, with $\ell, m = 1 \ldots N$ standing for rows and $i, j = 1 \ldots N$ standing for columns. The problem Hamiltonian in this case can be taken to be

$$H(\sigma_\ell) = H_P + H_C,$$

where H_P is the "penalty" Hamiltonian, which in this case should add a penalty every time a healthy student is situated next to an ill student, while H_C is a "constraint" Hamiltonian, which enforces that the number of ill students is equal to the number of healthy students.

For H_P, we can take

$$H_P(\sigma_\ell) = \Lambda \sum_{\ell,m=1}^{N} \sum_{i,j=1}^{N} (1 - \sigma_{\ell N+i}\sigma_{mN+j})$$
$$\times \left(\delta_{\ell m}(\delta_{(i+1)j} + \delta_{(i-1)j}) + \delta_{ij}(\delta_{(\ell+1)m} + \delta_{(\ell-1)m})\right),$$

where δ_{ab} is the Kronecker δ-symbol. In this Hamiltonian, the double-sum takes us around the lattice of nodes, comparing node ℓ, i to node m, j. We see that the first bracket in the sum gives zero unless the nodes have opposite signs; in other words, there can be no penalty unless a healthy student is sitting at one desk while an ill student is sitting at the other. However, we only want a penalty to be imposed when the nodes we are considering are in either adjacent rows or columns. Hence, the brackets of Kronecker δ-symbols can only contribute if either the students are in the same row ($\ell = m$) and adjacent columns ($j = i \pm 1$) or the same column ($i = j$) and adjacent rows ($\ell = m \pm 1$). Finally, the constraint is simple to enforce if we note that $\sum_{\ell,i=1}^{N} \sigma_{\ell N+i}$ is the number of ill students minus the number of healthy students. Thus, to enforce equality, we may add

$$H_C = \Lambda' \left(\sum_{\ell,i=1}^{N} \sigma_{\ell N+i} \right)^2 = \Lambda' \sum_{\ell,m=1}^{N} \sigma_{\ell N+i} \sigma_{mN+j}.$$

4.1.1 *Reduction*

An important aspect of solving problems on annealers by minimizing Hamiltonians of the generalized Ising form in Eq. (4.1) is *reduction*. To illustrate how the process of reduction works, it is instructive to consider a third kind of problem, the so-called "taxicab number" problem. Taxicab numbers are numbers that can be expressed in more than one way as sums of equal powers. The most famous example is the number of Hardy and Ramanujan's eponymous taxicab, $\mathrm{Ta}(2) = 1729$. This is the smallest of the following list of numbers, all of which are expressible as the sum of two cubes in two different ways:

$$\begin{aligned}
1729 &= 1^3 + 12^3 = 9^3 + 10^3, \\
4104 &= 9^3 + 15^3 = 16^3 + 2^3, \\
20683 &= 24^3 + 19^3 = 10^3 + 27^3, \\
32832 &= 32^3 + 4^3 = 18^3 + 30^3, \\
&\cdots
\end{aligned}$$

Enthusiasts who collect these things tend to use the notation (k, m, n) to refer to such numbers, where k is the power, while m and n are the number of terms on each side. Thus, $\mathrm{Ta}(2) = 1729$ is defined as the smallest $(3, 2, 2)$ number, while Fermat's theorem is the statement that $(k, 1, 2)$ numbers exist for $k = 2$ only. This is one of the Diophantine problems that was considered in the context of quantum annealing in Ref. [18].

Constructing a suitably efficient encoding for this kind of problem requires a significant degree of gymnastics because they involve a system of cubic and higher-order equations. The crux of the matter is therefore to implement a reduction procedure that can continue to represent the complete system as a single Hamiltonian that is at most quadratic. To do this, we can adapt a method that first appeared in Ref. [19] and then more recently in Refs. [17,18,20–22]. It is completely problem-independent and, therefore, potentially applicable to any set of polynomial equations.

It works as follows. We are interested in solving a set of polynomial Diophantine equations $f_A(t_i) = 0$, where A labels the equations and $t_i \in \mathbb{Z}$ are the would-be integer solutions to the problem of interest. These integers will be encoded in a binary format, that is, we will use the following encoding:

$$t_i = \tau_{i,0} + 2\tau_{i,1} + \cdots + 2^{\beta-1}\tau_{i,\beta-1},$$

where we use τ to denote the binary variables corresponding to a given spin,

$$\tau_{i,k} = \frac{1}{2}(1 + \sigma_{i,k}), \tag{4.2}$$

with $\tau_{i,k} \in \{0,1\}$. The Hamiltonian expressed in terms of the τ variables is said to be in the *QUBO form* (which stands for "Quadratic Unconstrained Binary Optimization", meaning a quadratic objective written solely in binary variables with any constraints absorbed into the cost function). Obviously, the highest integer that can be represented in this way is $2^\beta - 1$, so it is up to us to decide the size of the space we wish to explore by setting the number of logical qubits, β.

Now, *in principle*, a solution to our problem can be found by finding the minimum of the raw Hamiltonian

$$\tilde{H} \equiv \sum_A (f_A(t_i(\sigma_\ell)))^2. \tag{4.3}$$

For a Diophantine system containing order-d polynomials in t_i, the raw Hamiltonian $\tilde{H}$ in Eq. (4.3) is an order-$2d$ polynomial in the spins, σ_ℓ. For example, the Ta(2) problem yields an order-6 spin-polynomial. Therefore, we must transform $\tilde{H}$ into an equivalent quadratic problem-Hamiltonian, H, that is guaranteed to have the same minima but which can be implemented on the annealer based on the quadratic Ising model.

For this task, we shall use the reduction method described in the Appendix of Ref. [22]. This method works by introducing auxiliary spins, usually referred to as *ancilla* variables, to represent *pairs* of spins in the raw Hamiltonian of Eq. (4.3) and is one of the many methods in the comprehensive survey of Ref. [21].

To implement the method, we begin with the raw polynomial $\tilde{H}(\sigma_\ell)$ written as a function of binary variables using Eq. (4.2). Suppose $\tilde{H}$ has terms involving products of two binary variables τ_1 and τ_2. Now, consider adding to the polynomial $\tilde{H}$ a quadratic term that involves the binary variables, together with a new ancilla variable τ_{12}, which is of the form

$$Q(\tau_{12};\tau_1,\tau_2) = \Lambda(\tau_1\tau_2 - 2\tau_{12}(\tau_1+\tau_2) + 3\tau_{12}).$$

Inspection shows that a sufficiently large and positive overall coupling Λ enforces $\tau_{12} = \tau_1\tau_2$. Importantly, the minimum at this point has $Q = 0$. Therefore, we may replace the product $\tau_1\tau_2$ with the ancilla variable τ_{12} wherever it appears within $\tilde{H}$, and the new Hamiltonian is guaranteed to have the same set of minima as the original $\tilde{H}$. Therefore, the process can be iterated until one arrives at the desired problem-Hamiltonian which is quadratic in spins and which is schematically of the form

$$\begin{aligned} H = {} & \tilde{H}(\tau_1,\tau_2,\ldots,\tau_{12},\tau_{13},\ldots,\tau_{12,34},\tau_{12,56}\cdots) \\ & + \sum_{i>j} Q(\tau_{ij};\tau_i,\tau_j) + \sum_{i<j,k<m} Q(\tau_{ij,km};\tau_{ij},\tau_{km}) \\ & + \cdots, \end{aligned}$$

with the constraints imposed by the Q terms ensuring that this quadratic Hamiltonian has the same minima as the original order-$2d$ polynomial.

We can check that the reduction works correctly with the example order-3 Hamiltonian

$$\begin{aligned} \tilde{H} &= \sigma_1\sigma_2\sigma_3 \\ &\equiv 8\tau_1\tau_2\tau_3 - 4\tau_1\tau_2 - 4\tau_1\tau_3 - 4\tau_2\tau_3 \\ &\quad + 2\tau_1 + 2\tau_2 + 2\tau_3, \end{aligned}$$

where we drop the constant -1 in translating to the binary variables. This Hamiltonian has four minima at $\sigma_1\sigma_2\sigma_3 = -1$ (which corresponds in binary language to any one of the τ_ℓ being zero, or all of them), as opposed to the seven solutions to $\tau_1\tau_2\tau_3 = 0$. As described above, we can reduce the trilinear term by trading $\tau_1\tau_2$ for an auxiliary binary variable τ_{12} and adding the Hamiltonian $Q(\tau_{12};\tau_1,\tau_2)$. The quadratic problem-Hamiltonian (in QUBO language) is then

$$\begin{aligned} H = {} & Q(\tau_{12};\tau_1,\tau_2) + 8\tau_{12}\tau_3 \\ & - 4\tau_{12} - 4\tau_1\tau_3 - 4\tau_2\tau_3 \\ & + 2\tau_1 + 2\tau_2 + 2\tau_3. \end{aligned}$$

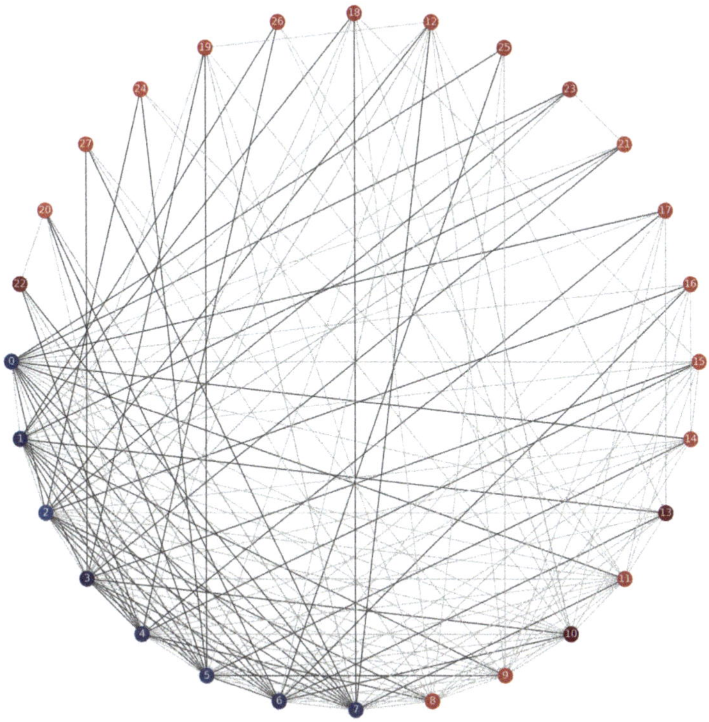

Fig. 4.2. Representation of $f(t_i) = (t_1^2 + t_2^2 - 149)^2$ as a quadratized Ising model. Nodes, corresponding to spins, can be arranged in a circle and are quadratically coupled by links. Weaker couplings are represented in light grey, gradually getting darker for higher coupling strengths. Linear couplings are coloured from dark blue (large negative terms) to dark red (large positive terms).

It is easy to verify that provided $\Lambda > 2$ the original four degenerate solutions hold in the new combined Hamiltonian as required.

Reduced Hamiltonians can become quite large depending on the problem at hand, but nevertheless they continue to yield solutions. It is often convenient to represent the final problem Hamiltonian in the graphical form using network graphs in which nodes correspond to spins and links to couplings. As an example, Figure 4.2 is a representation of the quadratized Hamiltonian associated with the Diophantine equation $f(t_i) = (t_1^2 + t_2^2 - 149)^2$. As can be seen even for this quite simple problem (which is a quartic order polynomial), the network is significant.

Returning now to the original problem of the taxicab numbers, let us consider the $(3, n, m)$ numbers, where $n, m \in \mathbb{N}^+$. For this problem, our raw Hamiltonian takes the form

$$\tilde{H} = \left(\sum_{i=1}^{n} (t_i)^3 - \sum_{i=1}^{m} (t'_i)^3 \right)^2 .$$

Note that one can add a constraint to the Hamiltonian to ensure that the two equations are not actually the same, i.e., $\{t_i\} \neq \{t'_i\}$ when $n = m$.

For such a constraint in the present case, we can simply impose a penalty whenever one of the integers in one set of polynomials coincides with an arbitrarily chosen integer in the other. In other words, we may add the penalty Hamiltonian

$$H_P = \Lambda \sum_{i=1}^{m} \left(\prod_{k=0}^{\beta} (1 - (\tau_{1,k} - \tau'_{i,k})^2) \right) .$$

We can see that this compares the first integer in the $f(t_i)$ equation with all the integers in the $f(t'_i)$ equation. It adds a term for each of the t'_i integers, and by inspection, we can see that (since the τ variables take values of only 0 or 1), this term vanishes if *any* of the $\tau'_{i,k}$ variables are different from the $\tau_{1,k}$ variables. If there is no difference, then one of the integers t'_i is equal to t_1, and the would-be solution is either trivial or it trivially reduces to a lower-order $(3, n-1, m-1)$ taxicab number: hence, H_P correctly gives a penalty of Λ. Although such a penalty can be implemented in practice, it is worth remarking that searching the generated set of proposed solutions is not an NP-hard problem, so it is sufficient to simply check at the end and discard the trivial solutions.

However, even if we do not include H_P, the system of sextic polynomials clearly presents an extremely large reduction problem. Fortunately, the reduction process can be automated in the manner described in Ref. [18].

4.1.2 *Practical quantum annealing*

Now that we have some grasp on how to encode problems, how do we actually go about performing the minimization of H? There are two methods one could contemplate for this, which are often confused: the first is the aforementioned quantum annealing utilized in a dissipative way, as in Refs. [22–24], and the second is adiabatic quantum computing (AQC), as envisaged in Refs. [25–27]. These two methods start from the same perspective — namely, we begin with a larger Hamiltonian on the annealer, which is of the form

$$\mathcal{H}_{QA} = B(s)\left(\sum_{\ell m} J_{\ell m}\sigma_\ell^Z\sigma_\ell^Z + C(t)\sum_\ell h_\ell\sigma_\ell^Z\right) + A(s)\sum_\ell \sigma_i^X,$$

where ℓ, m now label the physical qubits on the machine, with σ_ℓ^Z being the z-spin Pauli matrices and σ_ℓ^X being the so-called *transverse field* components, while the couplings h_ℓ and $J_{\ell m}$ between the qubits are set and kept constant. (It should be remarked that, typically, too large values of couplings will be automatically *autoscaled* to values below unity by the annealer, and thus one should choose all the couplings sufficiently small that autoscaling of J, h is avoided).

The parameter $s(t)$ (with t being time) is a user-defined control parameter that can be adjusted during the anneal, while A, B describe the resulting change in the characteristics of the annealer. For completeness, I also show $C(t)$, which is another user-defined parameter called the *h-gain* and is typically also provided on devices such as those of D-Wave [13].

To perform the task of finding the global minimum of a logical problem-Hamiltonian, H, we insert it into the quadratic Ising model part of $\mathcal{H}_{QA}$ represented by the B-terms, and we then adjust the relative parameters A, B. Often, the logical problem-Hamiltonian "B" part of $\mathcal{H}_{QA}$ is somewhat confusingly called the "classical Hamiltonian". For consistent confusion, I will also use this expression. Thus, the purpose of the annealer is to allow us to adjust A and B in order to dial the Hamiltonian from the pure transverse field Hamiltonian to the pure classical problem-Hamiltonian.

For this purpose, the functions $A(s)$ and $B(s)$ have a characteristic s-dependence of the form shown in Figure 4.3. When $s = 0$, the problem-Hamiltonian becomes (in a somewhat sloppy sense) "maximally quantum", by which I mean that significant hopping of its σ^Z spins is induced by the transverse field component. On the other hand, when $s = 1$, the system has arrived at the pure classically Ising-encoded problem-Hamiltonian, and the σ^Z spins become fixed. In a so-called *forward anneal*, we choose an *anneal schedule* that takes $s(t_i) = 0$ and $s(t_f) = 1$, beginning with rapidly fluctuating σ^Z spins and hopefully ending up with a system that solves the optimization problem of interest.

This is where I should briefly explain the moral difference between quantum annealing and AQC. AQC is the idealized situation in which the system begins in a ground state of the transverse field Hamiltonian (which is trivial to determine — consider it as an exercise) and remains

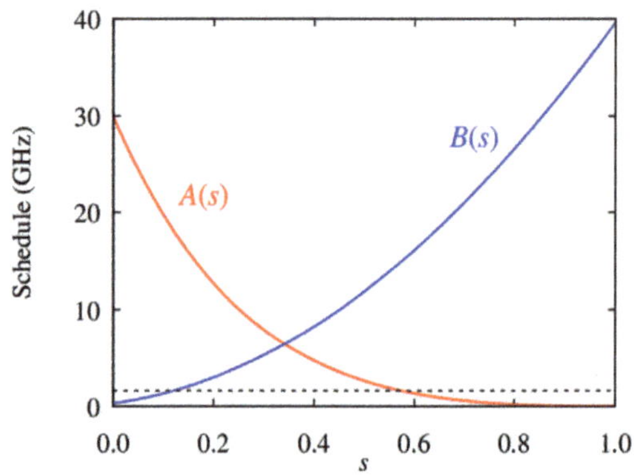

Fig. 4.3. Anneal schedule parameters. The thermal contribution is shown as a dotted line, while A and B are the coefficients scaling the classical Ising and transverse field contributions, respectively.

coherent throughout the anneal. The idea is then to *adiabatically* change the system such that (thanks to the adiabatic theorem) it is *always* in the ground state of the combined Hamiltonian $\mathscr{H}_{QA}$; therefore, at $t = t_f$, it is guaranteed to be in the ground state of the problem-Hamiltonian, H, which is what we would like to know. Quantum *diabatic* annealing, on the other hand, often takes a somewhat more pragmatic approach: in practice and over longer time-scales, the annealer is dissipative, and thus it will tend to lose energy (and coherence), allowing it to *tunnel* to the ground state and settle there. (For more discussion of this distinction, see Ref. [28].) It should be stressed that quantum tunnelling is still quantum even in a dissipative system, which may, as a whole, lose coherence. (The reasons for this are a somewhat technical aspect of quantum tunnelling — essentially, tunnelling through energetic barriers is a long-wavelength phenomenon which can be understood as quantum fluctuations around *large* solutions to the equations of motion called instantons. It is for this reason that it still makes sense to consider quantum tunnelling in a thermal environment — for example in the early universe.)

I should mention the possibility of *reverse annealing*. In this case, one sets the anneal schedule to be at $s = 1$ at both the initial and final times. During the anneal, the transverse field term is dialled up by sending s to some finite value s_q for some time interval before returning to the classical system. This option allows us to fix the system initially and is useful, for example, if we wish to set up the system in a metastable state and study the dynamics of its tunnelling (e.g., in Ref. [23]). The differences between

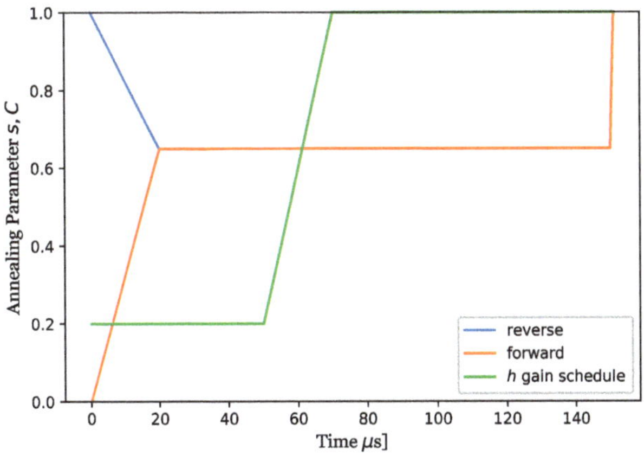

Fig. 4.4. A typical set of anneal schedules and h-gain schedule.

these two types of anneal schedule are shown in Figure 4.4. Finally, for completeness, it is also possible during the anneal to choose an *h-gain schedule*, $C(t)$, which can for example vary between $C_0 < 1$ and 1, as indicated by the green line in Figure 4.4. In tandem with reverse annealing, this can be useful if one wishes to start the system in the ground state of a particular problem-Hamiltonian and finish in the ground state of a *different* problem-Hamiltonian (rather than beginning always in the ground state of the transverse field Hamiltonian, which is the case in forward annealing). It can aid in, for example, solving a problem with several elements that are more effectively treated sequentially or again in the study of dynamics.

In practice then, the way one operates on the actual machine is to fill the matrix J and vector h and then send them to the `DwaveSampler`, specifying the required anneal schedule and number of reads the machine should make. The sampler returns a list of σ_ℓ^Z values that it has read off, ordered by the energy of the corresponding problem-Hamiltonian. Thus, if everything has gone according to plan, the leading entries in the list should be solutions to the problem at hand.

As you get used to operating with the actual annealer in the tutorials, you will come across a number of practical issues that need to be tackled. First and foremost is the fact that the annealer does not physically incorporate completely general $J_{\ell m}$ matrices. Therefore, for reasonably large problems (certainly for the ones we have discussed here), one has to use a so-called *minor-embedding* on the D-Wave QPU [29] due to this limited connectivity. A minor-embedding is a way of filling in for missing links by locking two or more qubits together into a single logical qubit using chain couplings. These couplings should ideally hold through the anneal but occasionally they break. (These *chain breaks* are also reported in the output of the annealer.) Thus, one should adjust the parameters so that such chain breaks happen rarely however.

Despite all of these hurdles both quantum and simulated annealing (which, as you will see, is also available as a D-Wave module) remain an interesting and effective method. To illustrate this, in Tables 4.1 and 4.2, we show a set of Ta$(4,3,3)$ and Ta$(3,8,8)$ numbers that were found using these methods with $\beta = 5$, taken from Ref. [18].

Table 4.1. List of $(4,3,3)$ numbers found using $\beta = 3,4,5$ and 10,000 reads per anneal run.

(4,3,3)	a	b	c	d	e	f
2673	3	6	6	7	2	4
16562	9	1	10	11	6	5
28593	2	13	2	9	6	12
35378	13	4	9	11	12	1
43218	11	13	2	14	7	7
54977	4	8	15	9	14	10
195122	21	5	2	9	13	20
324818	14	9	23	21	2	19
619337	28	8	5	7	26	20
847602	1	25	26	29	19	10
1071713	12	32	7	28	26	3
1178898	29	11	26	1	32	19
1328498	29	9	28	23	32	3

Table 4.2. $(3,8,8)$ solutions with $\beta = 5$. The reduction needs 320 auxiliary spins. Note that even the smallest $(3,8,8)$ number represents a solution in a search space of size $32^{16} \sim 10^{24}$.

(3,8,8)	a_1	a_2	a_3	a_4	a_5	a_6	a_7	a_8	b_1	b_2	b_3	b_4	b_5	b_6	b_7	b_8
50139	1	3	6	10	12	20	23	30	2	5	9	13	17	19	25	27
73206	1	3	4	17	20	25	26	30	5	8	9	10	19	21	28	32
78202	3	4	17	18	19	24	27	30	1	2	9	16	20	22	28	32
85418	2	3	9	16	18	23	31	32	6	10	14	15	24	26	27	30

To comment on the efficacy of the method, the search space in the first case is of order $32^6 \sim 10^9$, while in the second it is $32^{16} \sim 10^{24}$. Yet in both cases, the annealer identified valid solutions after only $\mathcal{O}(10^5)$ reads.

4.1.3 *Conclusions*

Currently available quantum annealers can already solve a large variety of Diophantine problems at a relatively advanced level, but they are not yet competitive with their simulated counterparts. Nevertheless, we can be optimistic about their progress. The first reason why is to do with the non-universality of the Ising model, which is the main obstacle to performing higher-order tasks on a quantum annealer composed of *sub-graphs*. Indeed finding a suitable minor embedding of the problem-Hamiltonian is itself a non-trivial task. It is worth noting that deciding whether a graph G can be embedded in a graph H is itself an NP-complete problem (when H is arbitrary, but as in current annealers, G is built out of *Chimera* or *Pegasus* sub-graphs) [29]. We can anticipate improvement in embedding as the overall size of the annealer and the connectivity of its sub-graphs are continually increased.

The second reason to be optimistic about quantum annealers is in the potential speed-up in the way that they find global minima. For example, *diabatic* annealing has the potential to avoid anneal times increasing exponentially with the difficulty of the problem [30], an issue that is seen in both AQC and simulated annealing. These physical aspects of quantum annealing (which were also crucial in the quantum field theory tunnelling studies in Refs. [23,31,32]) cannot be efficiently simulated classically. Indeed, it has already been shown that quantum annealing overcomes simulated annealing in a large variety of cases (see for example Refs. [33–35]). It was, for example, found in Ref. [35] that quantum annealing can outperform simulated annealing by a factor of 10^8 in finding the minimum of a particular problem.

4.2 Tutorial

4.2.1 *Prerequisites*

This tutorial consists of three exercises: the first can be solved using pen and paper, and the last two are based on a Python notebook you can access here:

https://github.com/andrei-const/ML_MathsPhys_Oxford2023/blob/main/DWaveExample.ipynb

It is assumed you already have familiarity with Python and you are able to run Python scripts. In the following, we shall use *simulated annealing* instead of pure quantum annealing for the following reasons: it is straightforward to get your machine ready to use the D-Wave simulated annealing module; there is not any big difference, in term of coding, in the usage of the module for simulated annealing from the module designed for quantum annealing; to implement quantum annealing, you need to be able to access a physical quantum annealer via D-Wave at https://www.dwavesys.com/learn/quantum-computing/. (You get one free minute of anneal time upon registering, which, believe it or not, is a relatively long time).

Let us now proceed by installing the D-Wave module required for simulated annealing. At the time of writing, this can be done by running the following command in the terminal:

```
pip install dwave-neal
```

This assumes that pip is already installed on your system. Once the installation is complete, try running the following example in Python:

```
import neal

sampler = neal.SimulatedAnnealingSampler()
h = {'a': 0.0, 'b': 0.0, 'c': 0.0}
J = {('a', 'b'): 1.0, ('b', 'c'): 1.0}
sampleset = sampler.sample_ising(h, J, num_reads=10)

if sampleset.first.energy == -2.0:
    print("You are ready for the tutorial!")
else:
    print("Ops, something went wrong!")
```

If you have successfully installed everything you need for this tutorial, you should see "You are ready for the tutorial!".

To familiarize ourselves a bit more with this D-Wave module, let us now explain the instructions contained in the test code. After importing the required module (using `import neal`), we define the variable `sampler`, which corresponds to the specific D-Wave sampler designed for simulated annealing. Then, we define two dictionaries, `h` and `J`, which correspond to the single and quadratic Ising couplings, where "a", "b", and "c" are the names of the spins involved. We then call the sampler using the method `sample_ising` passing the value of the couplings. In doing this, we set the option `num_reads` to 10, which specifies the number of anneal runs. The results of the sampling are collected in `sampleset`. Indeed, using

```
print(sampleset)
```

we can directly print the content of that variable, which in this case is shown in Figure 4.5.

Then, using `sampleset.first.energy`, we can access the lowest-energy configuration found in the sampling and finally verify if the ground-state energy is the one expected.

It is worth mentioning that the Ising couplings can also be defined as `numpy` arrays. Here is the equivalent of the above example in terms of `numpy` arrays:

```
import numpy as np

h = np.array([0, 0, 0])
J = np.array([[0, 1, 1], [0, 0, 0], [0, 0, 0]])
```

```
   a  b  c energy num_oc.
0 +1 -1 +1   -2.0       1
1 +1 -1 +1   -2.0       1
2 -1 +1 -1   -2.0       1
3 -1 +1 -1   -2.0       1
5 -1 +1 -1   -2.0       1
6 -1 +1 -1   -2.0       1
7 -1 +1 -1   -2.0       1
8 -1 +1 -1   -2.0       1
9 +1 -1 +1   -2.0       1
4 +1 +1 -1    0.0       1
['SPIN', 10 rows, 10 samples, 3 variables]
```

Fig. 4.5. The results after 10 reads are sorted by the value of the corresponding energy, with nine configurations with the lowest energy (top) and one excited configuration (bottom).

Here, the positions (i,j) of the non-zero entries in J correspond to the coupling between spin i and spin j, with $i,j = 1,2,3$. Now that we have illustrated the basic usage of this module, we can move on to the main part of this tutorial.

4.2.2 *Optimal reduction*

This exercise is based on the Hamiltonian quadratization technique introduced in the lecture. Suppose we are given the Hamiltonian

$$H = \tau_1\tau_2\tau_3 \;-\; \tau_2\tau_3\tau_4 \;+\; \tau_1\tau_2\tau_3\tau_4,$$

where $\tau_i \in \{0,1\}$.

- What is the most convenient way to reduce this Hamiltonian while minimizing the number of auxiliary qubits?
- Write down the quadratized Hamiltonian. Check that both the location and the degeneracy of the minimum are preserved.
- How would you design a general algorithm that reduces an arbitrary higher-order Hamiltonian to a quadratic one using as few auxiliary qubits as possible?

Solution

The original Hamiltonian is

$$H = \tau_1\tau_2\tau_3 \;-\; \tau_2\tau_3\tau_4 \;+\; \tau_1\tau_2\tau_3\tau_4.$$

To quadratize it, we need to apply the reduction formula twice, each time collapsing a different pair of τ variables. As a first step, the most natural choice is to collapse the pair $\tau_2\tau_3$. After this, a trilinear term still remains, so we must collapse another pair; at this stage, more than one option is available. The final quadratized Hamiltonian takes the form

$$\begin{aligned} H' &= \tau_1 a_1 - a_1\tau_4 + a_1 a_2 \\ &\quad + \Lambda\big(\tau_2\tau_3 - 2a_1(\tau_2 + \tau_3) + 3a_1\big) \\ &\quad + \Lambda\big(\tau_1\tau_4 - 2a_2(\tau_1 + \tau_4) + 3a_2\big), \end{aligned}$$

where a_1, a_2 are the auxiliary binary variables introduced during the reduction.

The minimum of the original Hamiltonian H can be identified directly by inspection:

$$\tau_1 = 0, \quad \tau_2 = \tau_3 = \tau_4 = 1,$$

for which $H = -1$. This can also be verified explicitly by evaluating H for all 2^4 possible assignments of the τ variables. Performing the same check for H' confirms that both the location and degeneracy of the minimum are preserved.

Finally, regarding the design of a general algorithm, for a completely general higher-order Hamiltonian, it can be shown that finding the optimal set of variable pairs to collapse so as to minimize the number of auxiliary qubits is itself an NP-complete problem. Consequently, a practical approach is to use a greedy algorithm. In the example code, the reduction proceeds in several steps: at each step, the pair of variables appearing most frequently in the Hamiltonian is selected and replaced with an auxiliary variable. After updating the Hamiltonian with the corresponding penalty terms, the algorithm repeats this procedure until all remaining interactions are at most quadratic.

4.2.3 *Taxicab numbers*

Adapt the code from Section 4.2.1 to find the first few taxicab numbers,

$$a^3 + b^3 = c^3 + d^3,$$

with the requirement that a, b, c, d are all distinct. You will need to find a way to encode the constraint $a \neq c, d$ (or an equivalent restriction involving b).

Solution

Here is the modified version of the code used to search for taxicab numbers:

https://github.com/andrei-const/ML_MathsPhys_Oxford2023/blob/main/Solution-Taxicab.ipynb

Besides the obvious adjustments required — such as increasing the number of variables and updating details of their binary representation (all explained in the Python notebook) — the most important changes concern the optimization function and the implementation of the constraints.

We begin with the Hamiltonian

$$H(a,b,c,d) = \left(a^3 + b^3 - c^3 - d^3\right)^2.$$

However, this cost function has global minima whenever $(a,b) = (c,d)$ or $(a,b) = (d,c)$. These trivial minima must be lifted so that the optimizer only returns non-trivial solutions. To do this, we introduce an additional constraint Hamiltonian H_C that penalizes configurations with $a = c$ or $a = d$.

It is most convenient to write this constraint directly in terms of the binary variables $\tau_{i,k}$, where $i \in \{a,b,c,d\}$ labels the integers and $k = 0,\ldots,\beta-1$ labels the bits in their β-bit representation. A suitable choice is

$$\begin{aligned} H_C &\equiv H_\delta(a,c) + H_\delta(a,d) \\ &= \prod_{k=0}^{\beta-1}\left(1 - (\tau_{a,k} - \tau_{c,k})^2\right) + \prod_{k=0}^{\beta-1}\left(1 - (\tau_{a,k} - \tau_{d,k})^2\right). \end{aligned}$$

Explicitly,

$$H_C = \begin{cases} 0, & \text{when } a \neq c \text{ and } a \neq d, \\ 1, & \text{when } a = c \text{ and } a \neq d, \\ 1, & \text{when } a = d \text{ and } a \neq c, \\ 2, & \text{when } a = c = d. \end{cases}$$

The total Hamiltonian used in the optimization is therefore

$$\tilde{H} = H_D + H_C.$$

When written in terms of the τ variables, $\tilde{H}$ is a polynomial of degree 2β for $\beta \geq 3$. For instance, with $\beta = 5$, the automated reduction routine provided in the notebook reduces this expression to a quadratic Hamiltonian using 98 auxiliary spins.

4.2.4 *Continuous optimization problems*

Adapt the code to minimize the function

$$F(x) = x^4 - \frac{23}{2}x^3 + \frac{753}{16}x^2 - \frac{161}{2}x. \tag{4.4}$$

Note that the global minimum does not occur at an integer value of x.

Solution

To minimize this function, you must choose an appropriate binary encoding for the continuous variable x. In the proposed solution,

https://github.com/andrei-const/ML_MathsPhys_Oxford2023/blob/main/Solution-Optimisation.ipynb

we use a continuous binary expansion of the form

$$x = 2^{-l}\tau_{-l} + 2^{-l+1}\tau_{-l+1} + \cdots + 2^{0}\tau_{0} + \cdots + 2^{m-1}\tau_{m-1} + 2^{m}\tau_{m} = \sum_{i=-l}^{m} 2^{i}\tau_{i},$$

where l bits are used for the fractional part, $m+1$ bits are used for the integer part, and the total number of bits is $\beta = l + m + 1$.

As usual, the Hamiltonian to be minimized is simply the square of Eq. (4.4).

References

[1] A. B. Finilla, M. A. Gomez, C. Sebenik, and J. D. Doll, "Quantum annealing: A new method for minimizing multidimensional functions," *Chem. Phys. Lett.* **219**, 343, 1994.

[2] T. Kadowaki and H. Nishimori, "Quantum annealing in the transverse ising model," *Phys. Rev. E* **58**, 5355, 1998.

[3] J. Brooke, D. Bitko, T. F. Rosenbaum, and G. Aeppli, "Quantum annealing of a disordered magnet," *Science* **284**, 779, 1999.

[4] N. G. Dickson *et al.*, "Thermally assisted quantum annealing of a 16-qubit problem," *Nat. Commun.* **4**, 1903, 2013.

[5] T. Lanting, A. J. Przybysz, A. Y. Smirnov, F. M. Spedalieri, M. H. Amin, A. J. Berkley, R. Harris, F. Altomare, S. Boixo, P. Bunyk, N. Dickson, C. Enderud, J. P. Hilton, E. Hoskinson, M. W. Johnson, E. Ladizinsky, N. Ladizinsky, R. Neufeld, T. Oh, I. Perminov, C. Rich, M. C. Thom, E. Tolkacheva, S. Uchaikin, A. B. Wilson, and G. Rose, "Entanglement in a quantum annealing processor," *Phys. Rev. X* **4**, 021041, 2014.

[6] T. Albash, W. Vinci, A. Mishra, P. A. Warburton, and D. A. Lidar, "Consistency tests of classical and quantum models for a quantum annealer," *Phys. Rev. A* **91**, 2015.

[7] T. Albash and D. A. Lidar, "Adiabatic quantum computing," *Rev. Mod. Phys.* **90**, 2018. DOI: 10.1103/RevModPhys.90.015002.

[8] S. Boixo, V. N. Smelyanskiy, A. Shabani, S. V. Isakov, M. Dykman, V. S. Denchev, M. H. Amin, A. Y. Smirnov, M. Mohseni, and H. Neven, "Computational multiqubit tunnelling in programmable quantum annealers," *Nat. Commun.* **7**, 2016. DOI: 10.1038/ncomms10327.

[9] N. Chancellor, S. Szoke, W. Vinci, G. Aeppli, and P. A. Warburton, "Maximum–entropy inference with a programmable annealer," *Sci. Rept.* **6**, 2016. DOI: 10.1038/srep22318.

[10] M. Benedetti, J. Realpe-Gómez, R. Biswas, and A. Perdomo-Ortiz, "Estimation of effective temperatures in quantum annealers for sampling applications: A case study with possible applications in deep learning," *Phys. Rev. A* **94**, 022308, 2016.

[11] S. Muthukrishnan, T. Albash, and D. A. Lidar, "Tunneling and speedup in quantum optimization for permutation-symmetric problems," *Phys. Rev. X* **6**, 031010, 2016.

[12] A. Cervera Lierta, "Exact ising model simulation on a quantum computer," *Quantum* **2**, 114, 2018.

[13] T. Lanting, "The D-Wave 2000Q Processor," 2017, Presented at AQC 2017.

[14] N. Dattani, S. Szalay, and N. Chancellor, "Pegasus: The second connectivity graph for large-scale quantum annealing hardware," 2019. arXiv:1901.07636 [quant-ph].

[15] S. Tung, "Computational complexities of diophantine equations with parameters," *J. Algorithms* **8**, 324, 1987.

[16] N. Mohseni, P. McMahon, and T. Byrnes, "Ising machines as hardware solvers of combinatorial optimization problems," *Nat. Rev. Phys.*, **4**, 6, 363–379, 2022.

[17] P. Hauke, H. G. Katzgraber, W. Lechner, H. Nishimori, and W. D. Oliver, "Perspectives of quantum annealing: Methods and implementations," *Rept. Prog. Phys.* **83**, 054401, 2020.

[18] S. A. Abel and L. A. Nutricati, "Ising machines for diophantine problems in physics," *Fortsch. Phys.* **70**, 2200114, 2022. arXiv:2206.09956 [hep-th].

[19] I. G. Rosenberg, "Reduction of bivalent maximization to the quadratic case," *Cah. Cent. Etud. Rech. Oper.* **17**, 71–74, 1975.

[20] R. Tanburn, E. Okada, and N. Dattani, "Reducing multi-qubit interactions in adiabatic quantum computation without adding auxiliary qubits. Part 1: The "deduc-reduc" method and its application to quantum factorization of numbers," 2015.

[21] N. Dattani, "Quadratization in discrete optimization and quantum mechanics," 2019.

[22] S. Abel, J. C. Criado, and M. Spannowsky, "Completely quantum neural networks," 2022. arXiv:2202.11727 [quant-ph].

[23] S. Abel, A. Blance, and M. Spannowsky, "Quantum optimisation of complex systems with a quantum annealer," 2021. arXiv:2105.13945 [quant-ph].

[24] J. C. Criado and M. Spannowsky, "Qade: Solving differential equations on quantum annealers," 2022. arXiv:2204.03657 [quant-ph].

[25] E. Farhi, J. Goldstone, S. Gutmann, and M. Sipser, "Quantum computation by adiabatic evolution," 2000. arXiv:quant-ph/0001106.

[26] E. Farhi, J. Goldstone, S. Gutmann, J. Lapan, A. Lundgren, and D. Preda, "A quantum adiabatic evolution algorithm applied to random instances of an np-complete problem," *Science* **292**, 472, 2001. http://science.sciencemag.org/content/292/5516/472.full.pdf.

[27] E. Farhi, J. Goldstone, and S. Gutmann, "Quantum adiabatic evolution algorithms versus simulated annealing," 2002. arXiv:quant-ph/0201031.

[28] S. Abel, J. C. Criado, and M. Spannowsky, "Training neural networks with universal adiabatic quantum computing," 2023. arXiv:2308.13028 [quant-ph].

[29] E. Lobe and A. Lutz, "Minor embedding in broken chimera and pegasus graphs is np-complete," 2021.

[30] E. Crosson and D. Lidar, "Prospects for quantum enhancement with diabatic quantum annealing," *Nat. Rev. Phys.* **3**, 466, 2021.

[31] S. Abel, N. Chancellor, and M. Spannowsky, "Quantum computing for quantum tunneling," *Phys. Rev. D* **103**, 016008, 2021. arXiv:2003.07374 [hep-ph].

[32] S. Abel and M. Spannowsky, "Quantum-field-theoretic simulation platform for observing the fate of the false vacuum," *PRX Quantum* **2**, 010349, 2021. arXiv:2006.06003 [hep-th].

[33] G. E. Santoro, R. Martonák, E. Tosatti, and R. Car, "Theory of quantum annealing of an ising spin glass," *Science* **295**, 2427, 2002.

[34] R. Martoňák, G. E. Santoro, and E. Tosatti, "Quantum annealing of the traveling-salesman problem," *Phys. Rev. E* **70**, 057701, 2004.

[35] V. S. Denchev, S. Boixo, S. V. Isakov, N. Ding, R. Babbush, V. Smelyanskiy, J. Martinis, and H. Neven, "What is the computational value of finite-range tunneling?" *Phys. Rev. X* **6**, 031015, 2016.

PART 3

STRING THEORY COMPACTIFICATIONS, CALABI–YAU MANIFOLDS AND RICCI-FLAT METRICS

We explore applications of machine learning to understanding the geometry of Calabi–Yau manifolds. We begin with an overview of the necessary geometric background. We then go on to describe some conventional techniques which have historically been used to approximate the Ricci-flat metric on Calabi–Yau threefolds. Next, we describe the application of machine learning in this context. We show how such techniques are used to both augment and replace the previously existing methods, allowing us to obtain qualitatively different results.

CHAPTER 5

STRING THEORY COMPACTIFICATIONS, CALABI–YAU MANIFOLDS, AND RICCI-FLAT METRICS

LARA B. ANDERSON, JAMES GRAY, AND MAGDALENA LARFORS

Calabi–Yau (CY) manifolds play a ubiquitous role in string theory. As a supersymmetry-preserving choice for the six extra compact dimensions of superstring compactifications, these spaces provide an arena in which to explore the rich interplay between physics and geometry. These lectures focus on compact CY manifolds and the long-standing problem of determining their Ricci-flat metrics. Despite powerful existence theorems, no analytic expressions for these metrics are known.

In this lecture series, we review numerical approximation methods for Ricci-flat CY metrics. Our first aim is to give a brief overview of the mathematical framework underlying CY geometry and the various metrics that CY manifolds admit. We then discuss the three types of numerical methods that have been developed to compute Ricci-flat CY metrics: Donaldson's algorithm [28], functional minimization methods [37], and machine learning (ML) methods [5,7,26,29,32,43,46,47]. Due to the limited space we have, this will not be a comprehensive review, but instead we hope to give a brief survey and illustrate the essential tools, key ideas, and implementations of this rapidly advancing field.

5.1 Lecture 1: Calabi–Yau Geometry

5.1.1 *Introduction*

The goal of this lecture is to set the stage for the central topic of these lectures: using ML to solve (or, more accurately, approximate the solution to) Einstein's equations for Ricci-flat metrics on spaces with non-trivial topology. In particular, we are interested in geometries known as *Calabi–Yau* manifolds and their metrics, which play a special role in string theory. In particular, this set of lectures is about solving partial differential equations in curved spaces. For this goal, we clearly need some notions of geometry, as well as a little motivation as to *why* this problem is interesting. So, to that end, let's begin by asking: "What is a CY manifold and what is special about its metric?"

A CY manifold is a compact Kähler manifold with a Ricci-flat metric (and thanks to Yau's theorem, this is equivalent to a Kähler manifold with vanishing first Chern class). The more formal definition is given in Section 5.1.3. At the moment though, these terms may not mean much to you and as a result, we must provide some background for our discussion. There are a great many well-written and useful introductions to CY geometry (see e.g., Refs. [8,17,33,34]), and these lectures will not attempt to replicate those efforts. Instead, our goal is to provide as quick and concise a survey as possible of the key features of these geometries, with an eye to what will matter most to us in an ML/computational approach.

Within the subject of string theory, determining a Ricci-flat metric on a CY manifold is a very big prize indeed. To understand why, we need to begin by noting that string theory is a theory of fundamental interactions that is formulated in more than the $3+1$ dimensions of space and time that we observe in our universe — in fact, a lot more. In heterotic string theory for example, the underlying theory is defined in 10 dimensions (9 of space and 1 time-like dimension). The low-energy effective theory can be determined by a Lagrangian of the form (see e.g., Refs. [22])

$$S_{10D} \sim \int \sqrt{-g} d^{10}x \left(R + \alpha F_{\mu\nu}F^{\mu\nu} + \cdots\right) \tag{5.1}$$

for gauge theory in higher dimensions, coupled to gravity. If you have not encountered metrics before, stay tuned for Section 5.1.2.1 later, but for now note that the metric g determines the volume form and all inner products/contractions in the integral above.

To reduce this to a four-dimensional theory that could potentially be relevant to our world, we could look for solutions of the form $\mathbb{R}^{1,3} \times X_6$, where X_6 is some *compact* manifold. With such a choice, one could imagine performing the multi-dimensional integral over X_6 and effectively "integrating out" the extra dimensions to obtain a four-dimensional Lagrangian theory:

$$S_{4D} \sim \int \sqrt{-g} d^4x \left(R + \tilde{\alpha} F_{\mu\nu} F^{\mu\nu} + \cdots\right)$$

The process of obtaining S_{4D} is referred to as a "string compactification". Which compact geometries X_6 are allowed is determined by the symmetries of (5.1) and its equations of motion. One such class of solutions are given by CY manifolds which admit Ricci-flat metrics (see Ref. [34]). Clearly, the form of such a Lagrangian depends strongly on X_6 and its metric, $g_{\mu\nu}$. A great many quantities that specify the physical theory are a function of this metric, including (but not limited to)

- the cosmological background of the theory (i.e., cosmological constant, etc),
- masses of fields/particles,
- couplings of fields,
- gauge couplings.

Unfortunately, the analytic form of the relevant Ricci-flat metrics for compact spaces is not known.[1] For 30 years, string theorists have devised clever ways of working around our ignorance of the fundamental geometry and managed to extract many interesting properties of string compactifications. However, in order to fully understand the effective physics of a string compactification to four dimensions (and decide what it is useful for!), the question of X_6 and its metric must be addressed.

This is a difficult problem mathematically (as we will discuss later, a Fields Medal was awarded to Yau for proving that a CY metric exists, but the proof is not constructive). Since analytic tools to generate the metric are sparse, it is natural to turn to numeric approaches. Some progress has been made in this area over a number of years (see, e.g., Refs. [12, 13, 24, 25, 28, 37]), but a substantial new tool has arisen with ML algorithms. This has enabled some very exciting recent progress [5, 7, 26, 29, 32, 43, 46, 47],

[1] For non-compact CY manifolds, there are analytic expressions for Ricci-flat metrics. The conifold metric is one famous example; see Ref. [19].

and this is what we hope to give you a feel for in these lectures. We turn now to a brief review of what type of "geometry" must be specified about X_6 in the cases of interest (CY manifolds) and what is meant by a metric in this case.

5.1.2 *Manifolds in a nutshell*

Here, we provide a lightning review of a few tools from differential geometry related to manifolds and metrics. For a more detailed introduction to these topics, we suggest looking at Refs. [17,40] for CY manifolds in particular.

We begin with the basic definition of a manifold.

Definition 5.1.1. A real manifold, M, consists of the following information:

- A collection of open sets $\{U_\alpha\}$ such that $\cup_\alpha U_\alpha = M$.
- For each U_α, there exists a morphism $\phi_\alpha : U_\alpha \to \mathbb{R}^n$.
- If $U_\alpha \cap U_\beta \neq 0$, then $\phi_\alpha|_{U_\alpha \cap U_\beta} \circ \phi_\beta^{-1}|_{U_\alpha \cap U_\beta}$ is a diffeomorphism.

The collection $\{U_\alpha, \phi_\alpha\}$ is called an *atlas*.

Note that complex manifolds are defined in a very similar way, except that we require mappings $\mathbf{z}_\alpha : U_\alpha \to \mathbb{C}^n$, and $\mathbf{z}_a \circ \mathbf{z}_\mathbf{b}^{-1}$ is a *holomorphic* function (i.e., a function of z and not its conjugate) on $\mathbf{z}_b(U_\alpha \cap U_\beta) \subset \mathbb{C}^n$ for all a, b. The mapping itself, $\mathbf{z}_a = (z_a^1, \ldots, z_a^n)$, is referred to as local holomorphic coordinates on the complex manifold, and on patch overlaps, $U_a \cap U_b$, the transition functions $z_a^i = f_{ab}^i(z_b)$ are holomorphic. An atlas with these properties is referred to as a *complex structure* on M. It is clear that every n-dimensional complex manifold is a $2n$-dimensional real manifold, but the converse is not true. For example, the spheres S^{2n} are only complex for $n = 1$.

In the context of string theory, we are frequently interested in compact, complex manifolds. It turns out that this first property of compactness can present challenges in the complex setting. It is straightforward to build a compact, algebraic submanifold of $\mathbb{R}^n$. For example, the algebraic constraint

$$x^2 + y^2 = 1 \subset \mathbb{R}^2$$

is a circle and clearly defines a compact, 1-dimensional submanifold of $\mathbb{R}^2$. But in $\mathbb{C}^n$, this turns out to be more difficult. In fact, $\mathbb{C}^n$ has no compact, complex submanifolds. This is because any global holomorphic function on a compact, complex manifold is a constant. Applying this to the coordinate functions yields the issue.

So, we are forced to choose a different complex "ambient space" in which to build the compact, complex manifolds of interest. For this purpose, it is helpful to consider a "compactified" version of $\mathbb{C}^n$, known as *complex projective space* and denoted $\mathbb{CP}^n$ (frequently in the physics literature, this is simply written as $\mathbb{P}^n$ with complexity assumed). Complex projective space is defined as the set of complex lines through the origin of $\mathbb{C}^{n+1}$.

Note that a line through the origin is defined by a single point, and points z, w define the same line if and only if there exists $\lambda \in \mathbb{C}^* = \mathbb{C} - \{0\}$ such that $(Z^0, Z^1, \ldots, Z^n) = \lambda(W^0, \ldots, W^n)$. We write this as

$$\mathbb{P}^n = \frac{\mathbb{C}^{n+1} - \{0\}}{\mathbb{C}^*} \quad \text{or} \quad \frac{S^{2n+1}}{U(1)}.$$

For $\mathbb{P}^n$, the coordinates $(Z^0, \ldots, Z^n)$ are called homogeneous coordinates and identified under the equivalence relation

$$(Z^0, Z^1, \ldots, Z^n) \sim \lambda(Z^0, \ldots, Z^n). \tag{5.2}$$

A good atlas can be obtained by the coordinate patches defined by

$$U_i = \{[Z] : Z^i \neq 0\}.$$

This is the set of lines through the origin that do not pass through Z_i. Natural local coordinates on each patch U_i are then $w_i^k = \frac{Z^k}{Z^i}$. In a later tutorial, you will play with these patches and prove to yourself that with the definitions above, the transition functions on overlaps are holomorphic functions.

It is worth mentioning at this stage that the setup above for complex projective space can very readily be generalized. For example, complex *weighted projective space* can be defined by introducing a scaling relation $(Z_0, \ldots, Z_n) \sim (\lambda_0 Z_0, \ldots, \lambda_n Z_n)$ instead of (5.2). More generally, there can be multiple $\mathbb{C}^*$ scaling relations of the coordinates and this leads to the beautiful framework of toric geometry. The interested reader can explore further here Ref. [23].

5.1.2.1 *Hermitian and Kähler Metrics*

With these basic definitions of manifolds in hand, we now turn to the object that is central to this series of lectures: the metric. Very briefly, the metric encodes information about lengths in (generally curved) spaces. For example, in three dimensions, instead of the line element expected from the Pythagorean theorem, dot products, etc. ($ds^2 = dx^2 + dy^2 + dz^2$), the notion of length in the space could be more complicated (i.e., $ds^2 = g_{AB}dx^A dx^B$), as illustrated schematically in Figure 5.1. In the context of physics, however, frequently metrics of interest are those on spacetime geometries, $g_{\mu\nu}$, where $\mu, \nu = 0, \ldots n$ with signature $(-,+,\ldots,+)$. For the case of CY compactifications, however, as described above, we focus on the metric on the internal, compact directions of space (e.g., X_6) only (all six directions are space-like). We also focus this discussion on the complex setting developed above.

Definition 5.1.2. A *Hermitian metric* on a complex manifold, X, is a covariant tensor field $g_{i\bar{j}}dz^i \otimes d\bar{z}^{\bar{j}}$. In local coordinates, $g_{ij} = g_{\bar{i}\bar{j}} = 0$, $g_{j\bar{i}} = \overline{g_{i\bar{j}}}$ and $g_{i\bar{j}}V^i\bar{V}^{\bar{j}} \geq 0$ for all V^i (i.e., the inner product is positive semi-definite). It follows that this metric defines a (real) $(1,1)$ form,

$$J = i(g_{i\bar{j}}dz^i \wedge d\bar{z}^{\bar{j}}), \tag{5.3}$$

which is called the *Hermitian form*, or, sometimes, the fundamental form. Furthermore, if this form satisfies the constraint that it is closed (i.e., $dJ = 0$), then it is called a *Kähler form*, and the manifold X is called a Kähler manifold.

The condition $dJ = 0$ can be expanded as

$$dJ = i(\partial + \bar{\partial})g_{i\bar{j}}dz^i \wedge d\bar{z}^{\bar{j}} = 0, \tag{5.4}$$

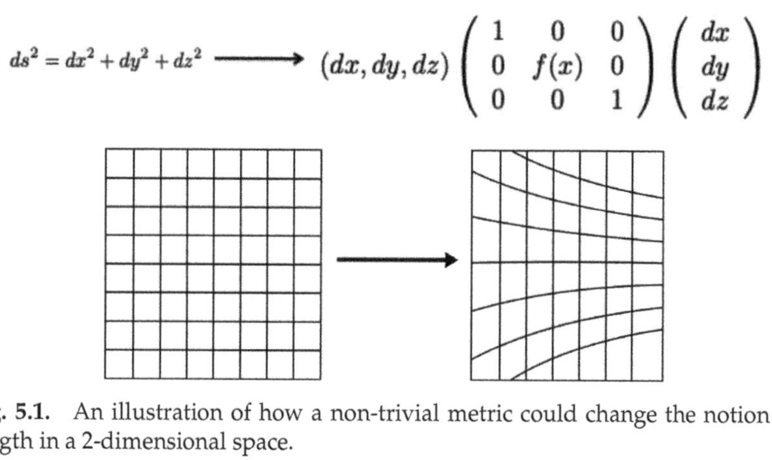

Fig. 5.1. An illustration of how a non-trivial metric could change the notion of length in a 2-dimensional space.

which implies that $\frac{\partial g_{i\bar{j}}}{\partial z^l} = \frac{\partial g_{l\bar{j}}}{\partial z^i}$, etc. This leads to the important fact that

$$g_{i\bar{j}} = \frac{\partial^2 K}{\partial z^i \partial \bar{z}^{\bar{j}}}, \quad g_{ij} = g_{\bar{i}\bar{j}} = 0.$$

That is, the Kähler metric is fully determined by the (locally defined) function $K = K(z, \bar{z})$, called the *Kähler potential.*

It is worth noting that many notable/simple complex manifolds are also Kähler. In particular, the complex projective space ($\mathbb{CP}^n$ defined above) is Kähler. This fact can be important for approaches to numerically constructing metrics, as we will see later in this lecture series. For now, we catalog several useful properties of Kähler manifolds (see, e.g., Ref. [17] for more details).

The volume form of an n-dimensional Kähler manifold can be expressed in terms of the Kähler form as

$$\frac{J^n}{n!} \sim \det(g) dz^1 \wedge \cdots dz^n \wedge d\bar{z}^1 \wedge \cdots \wedge d\bar{z}^n,$$

and the Riemann tensor takes the simple form

$$R_{i\bar{j}k\bar{l}} = g_{k\bar{l},i\bar{j}} = -g^{p\bar{q}} g_{k\bar{q},i} g_{p\bar{l},\bar{j}},$$

which in turn determines the Ricci tensor

$$R_{i\bar{j}} = -R_{i\bar{j}l\bar{k}} = -\partial_i \bar{\partial}_{\bar{j}} (\log(\det(g))),$$

and the first Chern class of the manifold X

$$c_1(X) = \frac{i}{2\pi} R_{k\bar{l}} dz^k \wedge d\bar{z}^{\bar{l}} = \frac{1}{2\pi} R = -\frac{i}{2\pi} \partial\bar{\partial} \log(\det(g_{k\bar{l}})).$$

With these definitions in place, we are ready to return to our discussion of Calabi–Yau manifolds.

5.1.3 *Calabi–Yau manifolds*

With the definitions above, we can now understand better the definition of a CY manifold: a complex, Kähler manifold with $c_1(X) = 0$. In 1978, by studying complex Monge–Ampere equations (see Section 5.1.3.1), Yau proved a conjecture by Calabi (posed in 1954) [57, 58]. This and other important work led to Yau's Fields Medal in 1982.

Theorem 5.1.3. *If X is a complex, Kähler manifold with $c_1(X) = 0$ and some Kähler form J, then there exists a unique Ricci-Flat metric on X whose Kähler form J' is in the same cohomology class as J (i.e., $J' = J + \partial\bar{\partial}\phi$).*

This definition is useful, but as Yau also noted in the proof of his famous theorem, we will see that packaging the condition of a Ricci-flat metric in other terms can simplify the problem. Much of the structure of a CY manifold is encoded in two canonical forms. The first of these, the Kähler form, we have encountered already in (5.3). The next is called the "holomorphic top form". An n-dimensional CY manifold admits a nowhere-vanishing, covariantly constant $(n,0)$ form, denoted by Ω. In the case of a CY 3-fold (our primary interest in these lectures), the holomorphic 3-form can be constructed from the single covariantly constant spinor, ϵ, which defines a manifold of $SU(3)$-holonomy [15]:

$$\Omega_{ijk} = \epsilon^T \gamma_{ijk} \epsilon,$$

with γ_{ijk} being the anti-symmetric product of γ-matrices satisfying

$$\{\gamma_i, \gamma_j\} = \{\gamma_{\bar{i}}, \gamma_{\bar{j}}\} = 0, \quad \{\gamma_i, \gamma_{\bar{j}}\} = 2g_{i\bar{j}}.$$

The fact that Ω is holomorphic (i.e., $\bar{\partial}_{\bar{j}} \Omega_{i_1,\ldots,i_n} = 0$) follows from Ω being covariantly constant.

The 3-form Ω is essentially unique. Given Ω, assume that there exists another 3-form Ω' with the same properties. Then, since top forms must be proportional $\Omega' = f\Omega$, with f being a non-singular function. But since $\bar{\partial}\Omega' = 0$ by assumption, it follows that f must be a holomorphic function. However, on a compact manifold, this implies that f is a constant. Later in this lecture, we will see an explicit form for the 3-form Ω for certain algebraic constructions of CY 3-folds.

5.1.3.1 *Kähler geometry and the Monge–Ampere equations*

The central goal of this lecture series is to use ML algorithms to attack the equation

$$R_{i\bar{j}} = -\partial_i \bar{\partial}_{\bar{j}} \log(\det(g)) = 0$$

with $g_{i\bar{j}} = \partial_i \bar{\partial}_{\bar{j}} K$ defined by a Kähler potential K. Observe that this is a fourth-order, highly nonlinear partial differential equation (PDE) for K in six (real) dimensions. As far as numerically solving PDEs goes, this is fairly awful. To make this tractable (both analytically and numerically), it is worth reformulating this problem by following the approach used by Yau in his famous proof: namely, to utilize the canonical differential forms (J, Ω) available on a CY manifold. Here is a sketch of the idea.

By the CY conjecture, start with *any* Kähler metric g on the CY 3-fold, X, and its associated Kähler form J_g. Then, the Ricci-flat metric, g_{CY}, and its Kähler form J can be written as

$$J = J_g + \partial\bar{\partial}\phi. \tag{5.5}$$

Observe then that there exist two ways of building a volume form on X:

$$J \wedge J \wedge J = \kappa \Omega \wedge \bar{\Omega} \quad (\kappa \in \mathbb{C}), \tag{5.6}$$

where the LHS of the equation above was defined in (5.3) and the RHS follows from the properties of Ω outlined above. Using (5.5), we are led to (5.6), a complex equation of *Monge–Ampere type* for ϕ, which is (only) second order. This equation was attacked using tools of differential geometry and analysis by Yau [57]. As we will see in subsequent lectures, it is also precisely this equation that provides us a handy foothold into machine implementation.

5.1.3.2 *Calabi–Yau moduli*

CY manifolds come equipped with rich and intricate moduli spaces. The word "moduli space" can refer not only to fluctuations of the metric, as we will see in what follows, but also to vacuum spaces of fields in the physical theory, and these two notions are deeply linked. For now, we focus on the geometric notion. See Refs. [34,39] for a deeper dive into these spaces.

If X is a CY 3-fold, then the Ricci-flatness condition guarantees that $R_{i\bar{j}}(g) = 0$. We can consider perturbing the metric (i.e., $g \to g + \delta g$) such that the Ricci-flatness condition is still preserved:

$$\delta g = \delta g_{ij} dz^i dz^j + \delta g_{i\bar{j}} dz^i d\bar{z}^{\bar{j}} + c.c.$$

Up to complex conjugation, these fluctuations come in two types (δg_{ji} and $\delta g_{i\bar{j}}$). The latter at least manifestly preserves the form of the initial Kähler metric. Plugging this fluctuation in to the Ricci-flatness condition leads to the constraint that $\delta g_{i\bar{j}}$ is harmonic. That is,

$$\delta g_{i\bar{j}} \in H^{1,1}(X).$$

Such deformations of the metric are known as *Kahler moduli* and are responsible for varying volumes (including the overall volume) within the CY threefold.

The second type of metric deformation, δg_{ij}, seems to make the metric non-Hermitian (as the index structure is different from that required in (5.3)). However, there exist coordinate redefinitions (see, e.g., Ref. [34]) which can change this back to Hermitian form (i.e., only $i\bar{j}$-type indices), but they are *not* holomorphic coordinate changes. As a result, the metric is Hermitian but with respect to a different complex structure from the one we began with. Thus, deformations of this type are referred to as *complex structure moduli*. It can be shown that

$$\Omega_{ijk} g^{k\bar{k}} \delta_{\bar{k}\bar{l}} dz^i \wedge dz^j \wedge d\bar{z}^{\bar{l}} \in H^{2,1}(X).$$

We will see in the following section that for simple algebraic constructions of CY manifolds (i.e., algebraic complete intersections), the complex structure deformations parameterized by $H^{2,1}(X)$ correspond to varying the coefficients in polynomial defining equations (i.e., changing the algebraic constraint that defines the manifold).

In summary, for a CY threefold, the two independent Hodge numbers ($h^{1,1}(X)$ and $h^{2,1}(X)$) count the number of independent metric fluctuations, δg, that preserve the CY condition.

5.1.4 *Building simple Calabi–Yau manifolds*

In this last bit of geometric introduction, we outline simple algebraic constructions of CY 3-folds. The idea is to utilize algebraic sub-varieties of simple ambient spaces (e.g., complex projective space, $\mathbb{P}^n$, defined above) to construct compact Kähler manifolds with $c_1(X) = 0$.

To begin, let us consider a single polynomial constraint in $\mathbb{P}^n$. We denote this hypersurface by $p(Z) = 0$, where Z_i are the homogeneous coordinates defined in (5.2). Note that in these coordinates, it is imperative that the polynomial $p(Z)$ be of homogeneous degree (every term in the polynomial must be of uniform scaling under the action in (5.2)); otherwise, the hypersurface is not well defined. For example, a polynomial constraint of the form $(Z^0)^2 + (Z^1) = 0$ in $\mathbb{P}^2$ does not yield a well-defined sub-variety (hint: what happens to this equation under the equivalence relation?), whereas, by contrast, $(Z^0)^2 + (Z^1)^2 = 0$ is well defined (since each term scales as λ^2) and yields a one-(complex)dimensional sub-space of $\mathbb{P}^2$ (in this case, a double cover of $\mathbb{P}^1$ itself).

It is straightforward to show that holomorphic, homogeneous polynomial constraints of this type yield Kähler sub-varieties of the ambient space. But what about the condition that $c_1(X) = 0$, as required by Theorem 5.1.3? It can be shown (and you will explore this in the tutorial later) that the CY condition can be simply enforced for manifolds of this type.

If we begin with a simple Kähler metric on $\mathbb{P}^n$, known as the Fubini–Study (FS) metric

$$g_{i\bar{j}} = \partial_i \bar{\partial}_{\bar{j}} \log(|Z^0|^2 + \cdots + |Z^n|^2), \tag{5.7}$$

then one can try to do the natural thing and restrict this to the sub-space, X, defined by $p(Z) = 0$. Suppose that the polynomial is of homogeneous degree m in $\mathbb{P}^n$. Then, a straightforward (but lengthy) curvature calculation yields

$$R_{\mathrm{FS}}|_X = ((n+1) - m) g_{\mathrm{FS}} + (\text{total derivative terms})$$

for the Ricci scalar. From this form, it is clear that

$$c_1(X) = 0 \quad \text{when} \quad m = n+1. \tag{5.8}$$

We denote the resulting CY n-fold, X, by the data of its ambient space and degree as $X = \mathbb{P}^n[m]$.

The prototypical example of this type is a degree 5 hypersurface in $\mathbb{P}^5$ (known as "the quintic" 3-fold, $X = \mathbb{P}^4[5]$). As an explicit example, consider this simple polynomial (known as the Fermat polynomial):

$$p = (Z^0)^5 + \cdots + (Z^4)^5 = 0.$$

This is far from the only choice however for such a degree-5 hypersurface. Indeed, we could choose different constant coefficients for every one of the possible $\binom{5+4}{4} = 126$ degree-5 terms in such a polynomial:

$$p = \alpha_0 (Z^0)^5 + \alpha_1 (Z^0)^4 (Z^1) + \cdots. \tag{5.9}$$

Changing the coefficients α_a corresponds to changing the complex structure (and hence deforming the metric), as in Section 5.1.3.2 described above (see, e.g., Ref. [33] for further details). However, not all 126 degrees of freedom above correspond to physical moduli. Indeed, there are 24 degrees of freedom corresponding to a general $PGL(5,\mathbb{C})$ coordinate transformation, as well as an overall scale (which drops out of equations such as (5.9)). The result leaves $126 - 25 = 101$ actual complex structure moduli for the quintic. In addition, there is a single Kähler modulus, inherited from the restriction of the Kähler form on $\mathbb{P}^n$ (associated with the metric in (5.7). As described in Section 5.1.3.2, the Hodge numbers of this CY manifold are then $(h^{1,1}(X), h^{2,1}(X)) = (1, 101)$.

There are many other manifolds that can be similarly constructed. For example, we could consider multiple defining equations in a single projective space (i.e., construct a complete intersection hypersurface). Using the same notation for ambient space and polynomial degree defined above, we can catalogue these so-called "configuration matrices" as

$$\mathbb{P}^4\,[5]\,,\quad \mathbb{P}^5\,[3,3]\,,\quad \mathbb{P}^5\,[4,2]\,,\quad \mathbb{P}^6\,[2,2,3]\,,\quad \mathbb{P}^7\,[2,2,2,2]\,.$$

These five manifolds (known as "cyclic" CY 3-folds) are the only CY 3-folds defined as algebraic sub-varieties of a single projective space. However, we could readily generalize our ambient space beyond a single projective space to, for example, products of projective spaces, allowing us to define manifolds such as the following in $\mathbb{P}^1 \times \mathbb{P}^3$:

$$\begin{matrix} \mathbb{P}^1_X \\ \mathbb{P}^3_Z \end{matrix} \begin{bmatrix} 2 \\ 4 \end{bmatrix}$$

$$p(z) = (X^0)^2(Z^0)^4 + (X^0)(X^1)(Z^0)^2(Z^1)(Z^2) + \cdots = 0.$$

More generally, to build a CY 3-fold within a product of p projective spaces, $\mathbb{P}^{n_1} \times \cdots \mathbb{P}^{n_p}$, we need

$$\sum_{r=1}^{p} n_r - L = 3,$$

where L is the number of defining equations, $p_j(z) = 0$, forming the complete intersection. Each of the homogeneous polynomials p_j which define the manifold can be characterized by its multi-degree $\mathbf{m}_j = (m_j^1, \ldots, m_j^m)$, where m_j^r specifies the degree of p_j in the homogeneous coordinates $\mathbf{z}^{(r)}$ of the factor $\mathbb{P}^{n_r}$ in $\mathscr{A}$. A standard way to encode this information is by a so-called *configuration matrix*:

$$\left[\begin{array}{c|cccc} \mathbb{P}^{n_1} & m_1^1 & m_2^1 & \ldots & m_K^1 \\ \mathbb{P}^{n_2} & m_1^2 & m_2^2 & \ldots & m_K^2 \\ \vdots & \vdots & \vdots & \ddots & \vdots \\ \mathbb{P}^{n_p} & m_1^p & m_2^p & \ldots & m_K^p \end{array}\right].$$

Note that the jth column of this matrix contains the multi-degree of the polynomial p_j. In order for the resulting manifold be CY, the condition $\sum_{j=1}^{L} m_j^r = n_r + 1, \qquad \forall r = 1, \ldots, p$ must be imposed, similarly to (5.8) so that $c_1(X)$ vanishes. There are over 7,000 such CY 3-folds that can be built in this way [20] (see also Ref. [31] for similar constructions of CY 4-folds).

By generalizing the ambient spaces still further to more general toric varieties, one can build CY hypersurfaces or more general complete

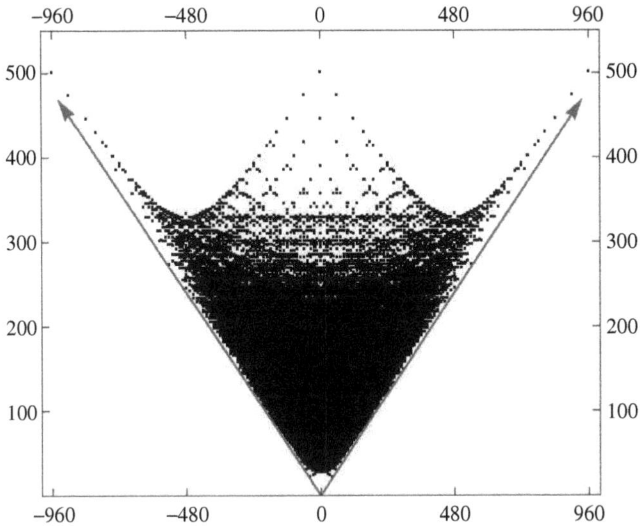

Fig. 5.2. A plot of Hodge numbers arising within the Kreuzer–Skarke dataset of Calabi–Yau threefolds (i.e., anti-canonical hypersurfaces in toric varieties [44]). The horizontal axis denotes $(h^{1,1} - h^{2,1})$, while the vertical axis is $(h^{1,1} + h^{2,1})$.

intersections. The most famous dataset of such manifolds is known as the Kreuzer–Skarke (KS) dataset [44] and consists of more than half a billion CY threefolds. The topology of known CY manifolds can vary widely. For example, the independent Hodge numbers described in Section 5.1.3.2 can range from $(h^{1,1}, h^{2,1}) = (1,1)$ to $(h^{1,1}, h^{2,1}) = (11, 492)$. A representation of the Hodge numbers found in the KS dataset is shown in Figure 5.2.

5.1.5 *The holomorphic top form in complete intersection Calabi–Yau manifolds*

In this section, we introduce a crucial tool: the form of the holomorphic 3-form, Ω in a complete intersection Calabi–Yau (CICY) threefold. This explicit form will prove crucial in numerically implementing the Monge–Ampere equations of (5.6), as we will see in later lectures.

On a CY threefold, the $(3,0)$ form, Ω, is essentially unique. As a result, in a given construction of the manifold, it may be possible to directly "spot" a covariantly constant, nowhere-vanishing form. Once found, this can yield new insights. The formulation reviewed here is well known [18,20,49,55] for the CICY threefold dataset here [20]. We explore this first for the case of a single hypersurface in $\mathbb{P}^n$.

Claim: Let X be given by $p(z) = 0$ in $\mathbb{P}^n$. Suppose we work on a coordinate patch with $z^i = 1$ in ambient (homogeneous) coordinates. Then,

$$\Omega = \int_\gamma \frac{\mu}{p}, \tag{5.10}$$

with

$$\mu = \sum_{j=0}^{n} (-1)^j \omega_j z^j dz^0 \wedge \cdots \wedge \hat{dz^i} \wedge \cdots \wedge dz^n,$$

where the term under the hat symbol is omitted and the contour γ surrounds $p = 0$ inside $\mathbb{P}^n$. For further details, we direct the reader to Ref. [20].

Here is an intuitive feel for why this formula is correct. Note that the form μ scales as $\mu \to \lambda^{n+1}\mu$ under the coordinate scaling of $\mathbb{P}^n$, but there exists a natural object that scales the same way. Namely, the defining equation $p(z)$ itself! As a result, the 3-form given in (5.10) is manifestly holomorphic and can be shown to be nowhere vanishing. As a result, it is the canonical 3-form we are after.

5.1.6 *Appendix to Lecture 1: Some useful pieces*

We recall here two useful results that will be important in the following sections.

Lemma 5.1.4. *The Trace Lemma:*

$$\partial \log(\det(A)) = \mathrm{tr}(A^{-1}\partial A).$$

Theorem 5.1.5. *The Global $\partial\bar{\partial}$-lemma: Let M be a Kähler manifold and $\phi \in H^{1,1}(M)$ $(\cap H^2(M, \mathbb{R}))$, then ϕ is exact if and only if there exists a function $f \in C^\infty(M, \mathbb{R})$ such that $\phi = \sqrt{-1}\partial\bar{\partial}f$.*

5.2 Lectures 2 and 3: Ricci-Flat Metrics from Conventional Numerical Techniques

In this section, we review two of the conventional techniques that have historically been used to obtain numerical approximations to Ricci-flat metrics on CY manifolds (for an early example of a different approach, see Ref. [41]). These approaches, while distinct, do share some common features. In particular, they both follow a two step process:

(1) An *ansatz* is made for the metric. More precisely, since these are Kähler metrics, an *ansatz* is made for the Kähler potential, a double derivative

of which then returns that metric. This *ansatz* contains some parameters, which we will denote as h in what follows.

(2) A method is then used to adjust the parameters h to get as close as possible to the Ricci-flat metric. How close a metric is to this desired result is measured by an error measure, a functional of the metric which vanishes only for the Ricci-flat instance.

It is useful to go through both of these steps in some detail here because much of the setup that is required in doing so is exactly what is required for the ML application that will be discussed in later sections.

In the following discussion, we will stick, for simplicity, to CY manifolds which can be described in the form $\left[\mathbb{P}^N | N+1\right]$. So, for example, we could consider

$$\begin{array}{rl} \text{K3} & \left[\mathbb{P}^3 | 4\right] \\ \text{the quintic threefold} & \left[\mathbb{P}^4 | 5\right] \\ \text{of the sextic fourfold} & \left[\mathbb{P}^5 | 6\right]. \end{array} \tag{5.11}$$

Much of this section follows standard discussions in the literature, and the author found Refs. [25,37] particularly useful in compiling it.[2]

5.2.1 *Metric ansatz*

Let us define the homogeneous coordinate on $\mathbb{P}^N$ to be Z^a, where $a = 1, \ldots, N+1$, with the usual associated open patches $Z^c \neq 0$ being denoted by U_c. We can then define the quantities $u_c^a = Z^a / Z^c$, which for $a \neq c$ are the usual affine coordinates on the patch U_c. The Kähler potential for the FS metric on $\mathbb{P}^N$ can be written, on patch U_c, in terms of these quantities as

$$K_c^{\text{FS}} := \ln\left(\frac{G_{a\bar{b}} Z^a \overline{Z^b}}{|Z^c|^2}\right) = \ln\left(G_{a\bar{b}} u_c^a \overline{u_c^b}\right). \tag{5.12}$$

In this expression, $G_{a\bar{b}}$ is a constant hermitian matrix, and one may obtain the metric via $g_{a\bar{b}}^{\text{FS}} = \partial_a \partial_{\bar{b}} K_c^{\text{FS}}$, where the derivatives are with respect to the affine coordinates.

[2] In addition, some very useful notes can be found here: https://people.brandeis.edu/~headrick/Mathematica/strategy.pdf.

An initial question a reader might have, given such an explicit metric on $\mathbb{P}^N$, is: can we not simply pull this back to get a metric on the CY manifold itself? More precisely, if we defined some coordinates y^i on the open patches of the CY manifold obtained by intersection with the U_c patches, then we would have the following metric:

$$g^{\text{CY}}_{i\bar{j}} = \frac{\partial u^a_c}{\partial y^i}\frac{\partial \overline{u}^{\bar{b}}_c}{\partial \overline{y}^{\bar{j}}} g^{\text{FS}}_{a\bar{b}}.$$

The metric g^{CY} is indeed perfectly well defined on the CY manifold; it is simply not the Ricci-flat metric (see Ref. [30] for an example of a concrete computation). Given this, our goal is to generalize g^{FS} to something more flexible, so we can obtain the Ricci-flat metric (or, at least, a good approximation to it) under the above pull back associated with the embedding map $y^i \to u^a_c(y^j)$.

One possibility would be to write down an *ansatz* for a Kähler potential that is given as an expansion in the following objects, which span the first k eigenspaces of the Laplacian on $\mathbb{P}^N$:

$$\frac{P^I_k(Z)\overline{P}^{\bar{J}}_k(\overline{Z})}{\left(G_{a\bar{b}}Z^a\overline{Z}^{\bar{b}}\right)^k}.$$

One problem with this is that these functions are linearly dependent on X. In particular,

$$P^I_k(Z) \sim P^I_k(Z) + \alpha_{k-N-1} P^{\text{defining}}_{N+1}, \tag{5.13}$$

where α is an arbitrary polynomial of the indicated degree and $P^{\text{defining}}_{N+1}$ is the defining polynomial of X. If we remove this redundancy and only include one representative of each equivalence class, which we shall denote by p^A_k, then we can write the following [53] Kähler potential:

$$\begin{aligned} K_{(c)} &= K^{\text{FS}}_{(c)} + \frac{1}{k}\ln\left(\frac{h_{A\overline{B}}p^A_k\overline{p}^{\overline{B}}_k}{(G_{a\bar{b}}Z^a\overline{Z}^{\bar{b}})^k}\right) \\ &= \frac{1}{k}\ln\left(h_{A\overline{B}}\frac{p^A_k\overline{p}^{\overline{B}}_k}{|Z^c|^{2k}}\right) \\ &= \frac{1}{k}\ln\left(h_{A\overline{B}}p^A_k(u_{(c)})\overline{p}^{\overline{B}}_k(\overline{u}_{(c)})\right). \end{aligned} \tag{5.14}$$

In this expression, h is an hermitian matrix of parameters (in order to return a real Kähler potential). The idea is to take this *ansatz* and vary the h to try and obtain a metric which, when pulled back to X, is a good approximation to the Ricci-flat metric on that manifold.

One question the reader might have is how likely is it that this attempt to approximate the Ricci-flat metric will succeed. The answer is furnished to us by the following theorem.

Theorem (Tian). *Let $\{p_k^A\}$ be a basis for $H^0(X, \mathscr{L}^k)$ for some ample line bundle $\mathscr{L}$. Then, the space of all algebraic Kähler potentials of the form* (5.14), *where $k \in \mathbb{Z}$ is dense in the space of Kähler potentials.*

For the case in hand, $\mathscr{L} = \mathcal{O}(1)$ so that our $p_k \in H^0(X, \mathcal{O}(k)) = H^0(X, \mathscr{L}^k)$. Thus, we see that if we use high enough polynomial degrees in forming our *ansatz* for $K_{(c)}$, then we can get arbitrarily close to any Kähler metric we desire, in particular the Ricci-flat one. Essentially, this result makes rigorous the idea that as we raise the polynomial degree k, there are more parameters appearing in our metric *ansatz* and thus more freedom to tune ourselves closer to the desired metric.

The above discussion leaves something to be desired in terms of geometric intuition as to the nature of the *ansatz* (5.14). In fact, it has a rather simple interpretation. Define a map $\mathrm{i} : X \to \mathbb{P}^{N_{p_k}-1}$, where $N_{p_k} = h^0(X, \mathcal{O}(k))$ is the number of independent p_k, as follows:

$$\mathrm{i}(Z^0, \ldots, Z^N) = (X_0 = p_k^1(Z), \ldots, X_{N_{p_k}} = p_k^{N_{p_k}}(Z)). \tag{5.15}$$

Here, the X_i are the homogeneous coordinates of $\mathbb{P}^{N_{p_k}-1}$. Note that this map respects the scaling of the projective spaces correctly, as a change $Z \to \lambda Z$ induces a change $p_k \to \lambda^k p_k$. Note also that in order to be a well-defined map, it must be the case that not all of the p_k can vanish simultaneously. This will certainly be the case for all of the examples discussed in this chapter.

To obtain an elegant geometric interpretation for our algebraic *ansatz* from this map, it is useful for it to be an embedding rather than simply an immersion. Mathematically, we obtain an embedding if L is quite ample, where

$$p_k^A \in H^0(X, L).$$

This condition holds for the quintic CY threefold in (5.11), for example, for all $k \geq 1$.

With these caveats aside, what is the geometric interpretation of the algebraic *ansatz* (5.14)? Well, given (5.15), we see that the Kähler potential (5.14) is simply the FS Kähler potential (5.12) on $\mathbb{P}^{N_{p_k}-1}$ restricted to X. Thus, the associated metric is simply the pullback of the FS metric on that large projective space.

How many free parameters does the Kähler potential *ansatz* (5.14) contain? Well, the number of degree k polynomials in the homogeneous coordinates of $\mathbb{P}^N$ is

$$N_k = \binom{k+N}{k}.$$

Therefore, after removing the redundancy (5.13), we find the following:

$$N_{p_k} = \begin{cases} \binom{k+N}{k} & \text{If } k \leq N, \\ \binom{k+N}{k} - N_{k-(N+1)} = \binom{k+N}{k} - \binom{k-1}{k-(N+1)} & \text{If } k > N. \end{cases}$$

This expression scales as k^{N-1} with the degree and dimension of the projective ambient space and thus increases quickly with degree. For the quintic, for example, the expression reduces to $\frac{5}{6}k(5+k^2)$, leading to the following values:

k	5	6	7	8	9
# polynomials p_k	125	205	315	460	645

It should be remembered that the parameters h form a matrix with each index running over these possibilities. Thus, we arrive at the following for the number of parameters in the algebraic *ansatz*:

k	5	6	7	8	9
# parameters $h_{A\bar{B}}$	15625	42025	99225	211600	416025

In almost all conventional applications (not employing ML techniques) to finding Ricci-flat metrics, the above numbers of parameters are simply too large to deal with. Thus, some simplification must be employed to decrease the number of degrees of freedom involved.

The most common method for decreasing the number of parameters that must be considered is to impose a symmetry on the CY manifold [16, 25, 35, 37]. We chose a special defining relation to ensure that such a symmetry exists and then restrict to the space of polynomials invariant under that symmetry. For example, consider the following defining relation for the quintic:

$$P_5^{\text{defining}} = \sum_{a=0}^{4} (Z^a)^5.$$

For this choice, we have the following symmetries which leave the defining relation invariant [25,37]:

- Permutation of the Z^a.
- The action $Z^a \to \omega^a Z^a$ where the ω^a are any fifth roots of unity (which can be different for different a's).
- Complex conjugation of the coordinates.

These actions give rise to a discrete symmetry group of order 150,000. Imposing this symmetry reduces the number of parameters, we must consider in the algebraic Kähler potential to $\mathcal{O}(10)$ at $k = 9$, to be compared with 416,025 in the above table [37]. It should be emphasized that the ability to avoid having to impose such symmetries is one of the areas, along with moduli dependence, where ML techniques have made qualitative improvements to the results which can be obtained.

As one last comment about the algebraic Kähler metric, let us talk about some possible choices of coordinates on the CY manifold and how to compute the necessary derivatives to obtain the metric (and its derivatives) efficiently. Locally, we could use any subset of $N-1$ out of our N affine coordinates u_c^a on the patch U_c of $\mathbb{P}^n$ as coordinates on X. Following the notation in Ref. [37], we can pick any $\delta \neq c$, and our CY coordinates are then

$$z^i = u_c^i \quad (i \neq c, \delta).$$

In such a parameterization, u_c^δ is a function of the z^i obtained by solving $P_5^{\text{defining}} = 0$.

Naively taking a derivative with respect to these coordinates is painful, as one must use the chain rule in an appropriate fashion and then substitute in an appropriate solution to the defining relation for u_c^δ. However, if dealt with correctly, such an arduous procedure can be avoided. Imagine taking the derivative with respect to the CY coordinates of some function f. In what follows, we shall use an index $\alpha : a|a \neq c$:

$$\frac{\partial}{\partial z^i} f(u_c^\alpha(z^i)) = \frac{\partial f}{\partial u^{\alpha_c}} \frac{\partial u^\alpha}{\partial z^i} = \frac{\partial f}{\partial u_c^i} + \frac{\partial f}{\partial u_c^\delta} \frac{\partial u^\delta{}_c}{\partial u_c^i}. \tag{5.16}$$

However, the defining relation is constant (vanishing) everywhere on the CY manifold, so that

$$\frac{\partial p^{\text{defining}}}{\partial z^i} = \frac{\partial p^{\text{defining}}}{\partial u_c^i} + \frac{\partial p^{\text{defining}}}{\partial u_c^\delta}\frac{\partial u_c^\delta}{\partial u_c^i} = 0.$$

We therefore find that

$$\frac{\partial u_c^\delta}{\partial u_c^i} = -\frac{\partial p^{\text{defining}}}{\partial u_c^i} / \frac{\partial p^{\text{defining}}}{\partial u_c^\delta} = -P_i^{\text{def}} / P_\delta^{\text{def}},$$

where we use the last equality to define P_i^{def} and P_δ^{def}. Using this result, we can write the following expression for (5.16):

$$\frac{\partial f}{\partial z^i} = \frac{\partial f}{\partial u_c^i} - \frac{P_i^{\text{def}}}{P_\delta^{\text{def}}}\frac{\partial f}{\partial u_c^\delta}.$$

This type of expression allows us to easily compute derivatives with respect to CY coordinates in terms of ambient space quantities. As described above, this is necessary, for example, to obtain the metric on the CY manifold itself.

5.2.2 *Tuning the parameters*

Given the algebraic *ansatz* for the Kähler potential discussed in the previous section, how will we fix the parameters in the *ansatz* to obtain Ricci-flat metrics? There were two commonly used techniques to be found in the literature before the advent of ML methodologies:

- The Donaldson algorithm: the iteration of a certain operator to a fixed point (see Ref. [25] for an early application in this context).
- The minimization of an appropriate functional [37].

We start with the latter possibility first here, as it is the method that generalizes more naturally to the context of the ML techniques discussed in the following section.

5.2.2.1 *Some basic definitions*

Let us start by defining some of the quantities that we will need in the subsequent discussion. If we denote the complex dimension of X by $n = N - 1$, then we define the following top form:

$$\mu = (-i)^n \Omega \wedge \overline{\Omega}.$$

The factor of i in that expression is included such that μ is real. Given a Kähler form J, we can then define the following scalar quantity:

$$v = \frac{J^n}{n!\mu}. \tag{5.17}$$

Since both the numerator and the denominator in the above are top forms, they are proportional to each other, and so the ratio v is well defined. Direct computation then reveals the following explicit formula:

$$v = \frac{\det g_{i\bar{j}}}{|\Omega_{1\ldots n}|^2}. \tag{5.18}$$

A useful property of the quantity thus defined is that the equation $v =$ constant (call it c) becomes the following:

$$\frac{1}{n!}J^n = c\Omega \wedge \overline{\Omega}(-i)^n.$$

This is precisely the Monge–Ampere equation. Finally, there is a simple relationship between v and the Ricci tensor for the metric associated with J:

$$R_{i\bar{j}} = -\partial_i \partial_{\bar{j}} \ln v. \tag{5.19}$$

It is useful to note in deriving this expression that, from (5.18),

$$\ln v = \ln \det g_{i\bar{j}} - \ln \Omega_{1\ldots n} - \ln \overline{\Omega}_{\bar{1}\ldots\bar{n}}.$$

5.2.2.2 *Energy functionals*

In forming functionals whose minimization would lead to a Ricci-flat metric on X, we might want something that is minimized either by solving the Monge–Ampere equation or by the Ricci-flatness condition. Let us start with the first of these, which is often called the Monge–Ampere functional:

$$H_{\mathrm{MA}} := \int_X \mu(v - < v >)^2. \tag{5.20}$$

Here,

$$< v > := \frac{\int_X \mu v}{\int_x \mu}$$

is simply the average of v over the CY manifold, computed using the measure μ. Through v's definition (5.17), this is a functional of J and thus, if evaluated with the *ansatz* (5.14), a function of the parameters h.

The functional $H_{\rm MA}$ has several desirable properties. It is positive semi-definite. The functional has a global minimum at $v =< v >$ which corresponds to

$$\frac{J^n}{n!} =< v > \Omega \wedge \overline{\Omega}(-i)^n.$$

In other words, its global minimum corresponds to solutions to the Monge–Ampere equation. The functional has no other critical points, as can be shown by direct computation. Lastly, $H_{\rm MA}$ can be computed without taking any derivatives of J. Since μ is fixed and known in closed form for the types of CY we are considering here [20] this makes the use of this functional quite efficient and simple!

Next, let us consider what we shall call the Ricci functional:

$$\begin{aligned} H_R &:= -\frac{1}{2}\int_X \mu R \\ &= \int_X \mu g^{i\bar{j}} \partial_i \partial_{\bar{j}} \ln v. \end{aligned} \tag{5.21}$$

Here, R is the Ricci scalar, and the second line follows from (5.19). We can rewrite this functional to make some of its properties clearer. Recall that, from (5.17),

$$\mu = \frac{J^n}{n!v} = \frac{\det g_{i\bar{j}}}{n!v}\epsilon^{(s)},$$

where $\epsilon^{(s)}$ is the anti-symmetric symbol and $\det g_{i\bar{j}} = \sqrt{g}$. Then, we can write the following:

$$\begin{aligned} H_R &= \int_X \frac{\sqrt{g}}{n!v}\epsilon^{(s)} g^{i\bar{j}} \partial_i \partial_{\bar{j}} \ln v \\ &= -\int_X \partial_{\bar{j}} \ln v \partial_i \left(g^{i\bar{j}} \sqrt{g} \frac{\epsilon^{(s)}}{n!v} \right) \\ &= -\int_X \epsilon^{(s)} \frac{\sqrt{g}}{n!} (\partial_{\bar{j}} \ln v) g^{i\bar{j}} \left(\nabla_i \frac{1}{v} \right) \\ &= \int_X \mu g^{i\bar{j}} (\partial_i \ln v)(\partial_{\bar{j}} \ln v). \end{aligned}$$

With the functional in this form, we can now see that it enjoys several useful properties. It is positive semi-definite. It has a global minimum when $v =$ constant, which, as we have already established, corresponds to the Ricci-flatness condition. The functional has no other critical points beyond these global minima, as can be determined by direct computation. However, this functional is a bit uglier and more difficult to use than H_{MA}. In particular, its use requires taking derivatives of J, which often leads to it being a less efficient choice.

5.2.2.3 *Integration and point sampling*

To evaluate the functionals presented in the previous section and indeed to use the Donaldson algorithm that we will present shortly, we need to be able to perform integrals over the CY manifold X. For the applications at hand, these integrals need to be carried out numerically, and so we briefly review some appropriate techniques here [25,48].

Let us start by considering the integration of a simple object where we can spot the answer. Consider an open set $U \subset X$. We can then define

$$\mathbb{1}_U(x) = \begin{cases} 1 \text{ if } x \in U, \\ 0 \text{ if } x \notin U. \end{cases}$$

It is then intuitively clear that

$$\int_X \mathbb{1}_U \mu = \mathrm{Vol}_\mu(U),$$

where $\mathrm{Vol}_\mu(U)$ is the volume of U *with respect to the measure* μ.

If we have a sample of N_p points $\{q_i \in X\}$ on X, which is *uniformly distributed according to* μ, then the expected number of points in $U \subset X$ is

$$\sum_{i=1}^{N_p} \mathbb{1}_U(q_i) \approx N_p \frac{\mathrm{Vol}_\mu(U)}{\mathrm{Vol}_\mu X}.$$

Rearranging this expression, we arrive at an approximate formula for the integral of $\mathbb{1}_U$:

$$\int_X \mathbb{1}_U \mu \approx \frac{\mathrm{Vol}_\mu(X)}{N_p} \sum_{i=1}^{N_p} \mathbb{1}_U(q_i).$$

If instead we wanted to integrate a different, arbitrary function, one might expect that one gets a similar result but with the points being weighted by the value of that function rather than by the $1/0$ weighting afforded by $\mathbb{1}_U$. This is indeed the case and we arrive at the following approximate formula for the integral:

$$\int_X f\mu \approx \frac{\mathrm{Vol}_\mu(X)}{N_p}\sum_{i=1}^{N_p} f(q_i). \tag{5.22}$$

This formula is rather easy to understand. The prefactor to the sum assigns a volume to each point in the sample (which is the same for every point due to the "uniformly distributed" assumption). One then simply multiplies this volume by the value of f at each point and sums to get the approximation to the integral. The statistical error in such an approximation is of order $1/\sqrt{N_p}$ [48].

In practice, the above approach to performing integrals over a CY manifold X encounters a problem in that producing samples of points distributed according to μ is not easy! Instead, one modifies the above approach in a two step process:

- First, one produces a set of points uniformly distributed according to some other measure μ' which is well adapted to such a task.
- Second, one then follows the above procedure but accounts for the fact that the points were not distributed according to μ by making use of what are called "mass functions".

Specifically, we define a mass function associated to some measure μ' by

$$m(x) = \frac{\mu(x)}{\mu'(x)}.$$

As in our definition of v above, such a ratio is well defined, as μ and μ' are both top forms and are thus proportional to one another. We then have the following for an approximation to the integral of some function f:

$$\int_x f\mu = \int_X f\frac{\mu}{\mu'}\mu' \approx \frac{\mathrm{Vol}_{\mu'}}{N_p}\sum_{i=1}^{N_p} f(q_i)m(q_i). \tag{5.23}$$

Here, in the approximate equality, we have used our earlier result (5.22). One can process this result further by noting that because

$$\int_X \mu = \mathrm{Vol}_\mu(X) = \int_X \frac{\mu}{\mu'}\mu' \approx \frac{\mathrm{Vol}_{\mu'}(X)}{N_p}\sum_{i=1}^{N_p} m(q_i),$$

we have the following result:

$$\frac{\text{Vol}_{\mu'}(X)}{N_p} \approx \frac{\text{Vol}_{\mu}(X)}{\sum_{i=1}^{N_p} m(q_i)}.$$

Substituting this result into (5.23), we arrive at the following expression [25]:

$$\int_X f\mu \approx \frac{\text{Vol}_{\mu}(X)}{\sum_{i=1}^{N_p} m(q_i)} \sum_{i=1}^{N_p} f(q_i)m(q_i).$$

The advantage of this expression is that it makes no mention of the auxiliary measure μ' that was used to form the point sample, outside of the mass function m.

What kind of point sampling techniques are used in the literature on approximate Ricci-flat metrics on CY manifolds? Here, we give two of the most common examples.

Rejection sampling

The first method we describe was used, for example, in Ref. [37]. Begin by defining a cover of the manifold X in terms of sets of points $\mathcal{O}_c$, where $\left|\frac{Z^\alpha}{Z^c}\right| \leq 1 \ \forall \ \alpha \neq c$. These coordinate patches are useful in numerical work because the affine coordinates associated with them are finite in value (less than or equal to one in magnitude). One then carries out the following steps:

- Choose points randomly in the patches $\mathcal{O}_c$ according to the coordinate measure.
- Throw away any points for which $|P_{N+1}^{\text{defining}}| \geq \epsilon$, where ϵ is some small number. In other words, we are only keeping those points which are already close to being on the CY. This situation is depicted schematically in Figure 5.3.
- Project the remaining points onto X orthogonally with respect to the FS metric to obtain the sample.

This point sampling method has a number of issues. First, it is wasteful in that one generates a large number of points far from the CY manifold that are then immediately discarded. This makes the method somewhat

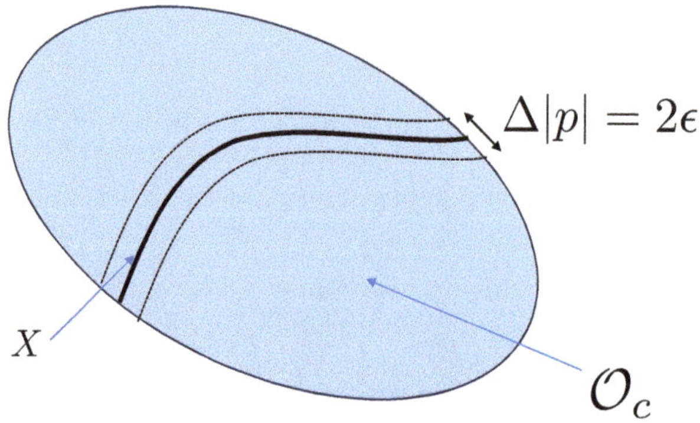

Fig. 5.3. A schematic depiction of some quantities appearing in rejection sampling, as described in the text.

slow compared to other possibilities. Second, it is hard to know the measure μ' with respect to which the points which are generated in this fashion are distributed. Still, the method is fast enough for most purposes, and in some respects the second issue mentioned above does not really matter so much. If we were to ignore the mass functions in this situation and just compute the integral, then we are really computing the integral

$$\int_X \mu'(v- < v >)^2$$

instead of

$$\int_X \mu(v- < v >)^2.$$

Both μ and μ' are simply some fixed measures. So, in some respects, both of these functionals are valid as quantities to minimize in order to find Ricci-flat metrics. That said, μ does have the advantage that it is proportional to $J \wedge J \wedge J$, the metric volume form, upon extremization of the functional. Thus, in situations where, for example, there are large physical volume regions associated with small coordinate volume regions on X, one can expect a better approximation from using μ than μ'. Nevertheless, in many situations, this subtlety is not important, and good results can be obtained with this simple point sampling technique [37].

Homogeneous sampling in projective space

The second method we describe for point sampling was used, for example, in Ref. [25]. The method is based upon starting with a uniform sample of pairs of points $\vec{Z}_{(1)}$ and $\vec{Z}_{(2)} \in \mathbb{P}^N$. Such pairs are easy to obtain. For example, one can perform the following two steps to find individual points on $\mathbb{P}^N$:

- Generate uniformly distributed points on S^{2N+1}. This can be achieved by using the Gaussian distribution (or any other desired spherically symmetric distribution)

$$\frac{1}{\sigma\sqrt{2\pi}} e^{-\frac{1}{2}\frac{1}{\sigma^2}\sum_a |Z^a|^2}$$

  and dividing by the norm.
- One can then obtain points on $\mathbb{P}^N$ by quotienting out by the $U(1)$ action $Z^i \to e^{i\theta} Z^i$.

Given pairs of such points, one can then proceed as follows to obtain a sample of points on X itself:

- Select pairs of points $\{\vec{Z}_{(1)}, \vec{Z}_{(2)}\}$ in projective space, as described above.
- Compute

$$P^{\text{defining}}_{N+1}(\vec{Z}_{(1)} + t\vec{Z}_{(2)}) = 0,$$

  and solve this to obtain all possible values for t. This finds for us the homogeneous coordinates, $\vec{Z}_{(1)} + t\vec{Z}_{(2)}$ with those values of t substituted in, of the $N+1$ points on the CY manifold X which are the intersections of that hypersurface with the straight line going through the points with coordinates $\vec{Z}_{(1)}$ and $\vec{Z}_{(2)}$. This situation is depicted in Figure 5.4.

This method of point sampling has a number of advantages. First, it is not as wasteful as rejection sampling, leading it to be an order of magnitude or so faster in practical implementations. More importantly, perhaps, the resulting points that are obtained in this way on X are distributed with respect to a **known** measure, the FS one (see Ref. [25] for a proof of this using original results by Shiffman and Zelditch [50, 51]). This means we can explicitly compute mass functions in this case.

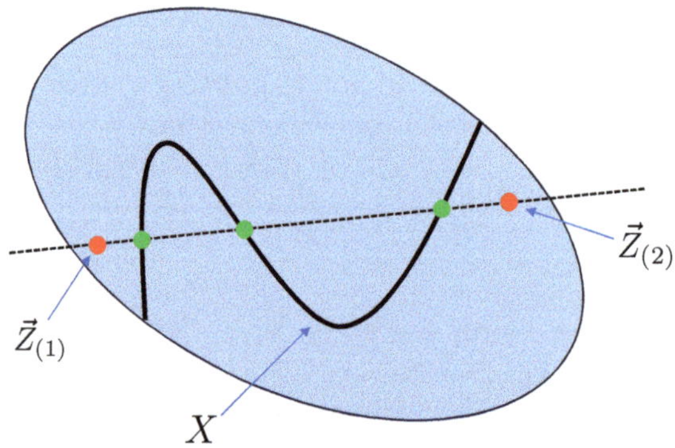

Fig. 5.4. A schematic depiction of some quantities appearing in the point sampling method based upon homogeneous sampling in projective space, as described in the text.

5.2.2.4 *Putting it all together*

Finally, we have everything that we need to obtain an approximate Ricci-flat metric on a Calabi–Yau manifold X using the functional minimization approach. Here is the procedure:

- Choose a polynomial degree k, and make a metric *ansatz* of the form (5.14) with associated parameters h.
- Choose a functional you wish to minimize.
- Use numeric integration to compute the functional as a function of the h.
- Minimize with respect to the parameters h. In the examples considered in the literature, in the case where symmetries are imposed on X, this step has been sufficiently straightforward that it can be completed using Mathematica's "FindMinimum" command [37].
- Plug the minimizing h's back into the *ansatz* to obtain the approximate Ricci-flat metric.

5.2.2.5 *Another approach to tuning the parameters: The Donaldson algorithm*

Let us end this section by briefly describing another commonly used approach to tuning the parameters h in the *ansatz* (5.14) to get close to a Ricci-flat metric. To use this method, called Donaldson's algorithm

[2,3,12,24,25,27,28], we first define a "T-Operator" as follows:

$$T_k(h)^{A\overline{B}} := \frac{N_{p_k}}{\int_X \mu} \int_X \frac{p_k^A \overline{p}_k^{\overline{B}}}{h_{C\overline{D}} p_k^X \overline{p}_k^{\overline{D}}} \mu.$$

Given a CY manifold, this operator takes in a matrix of parameters h and outputs a new matrix T. This operator exhibits the following property.

Theorem (Donaldson). *For any initial $h_{A\overline{B}}$ and fixed k, the sequence*

$$h_{m+1} = (T_k(h_m))^{-1}$$

converges to a unique "balanced" value as $n \to \infty$ and the associated metric on X is called the "balanced metric".

In practice, this iteration converges in $\mathcal{O}(10)$ steps for the types of CY manifolds we are considering in this chapter. This is useful because of the following theorem.

Theorem (Donaldson). *As $k \to \infty$, the balanced metric on X converges to the unique Ricci-flat metric for a given Kähler class and complex structure.*

More generally, one can consider non-CY examples and the balanced metric converges to a Kähler metric associated with the Kähler class $c_1(\mathcal{L})$ where the p_k are sections of $\mathcal{L}$. This will be a constant scalar curvature metric with curvature determined by $c_1(X)$.

A natural question is why we have not focused so much on this method of tuning the parameters in this section. Here are two of the key reasons:

- This method is not as directly relevant for most of the ML approaches to finding Ricci-flat metrics as the functional minimization method.
- This method tends not to do as well as the functional approach. One reason for this is simply that, while it is true that the balanced metric will converge to the Ricci-flat metric as $k \to \infty$, that does *not* mean that for any fixed finite k, the balanced metric is the best approximation to that desired result. Indeed, it has been shown that this is not the case [37]. Thus, minimizing a functional tends to result in a more accurate metric than iterating the T-operator, even while using the same metric *ansatz* with the same degree polynomials in it.

5.2.3 *Performance of and problems with conventional numerical methods*

In terms of performance, what diagnostics do we have to see how close we came to obtaining the Ricci-flat metric on our CY manifold X? There are a plethora of quantities we can use in such an evaluation. These include the value of both of the functionals discussed in this section, the value of v on a random of selection of points on X (where v should be constant for the Ricci-flat metric) or the Ricci scalar evaluated on points. How accurate a metric we obtain will of course depend on k, the degree of the polynomials we use in the algebraic Kähler potential *ansatz* (5.14). To give an idea of the type of accuracy that has been obtained with conventional techniques, in Ref. [37], the functional minimization approach was introduced and was used to obtain values of order $H_{\text{MA}} \sim 0.07 \times 5^{-k}$. So, we get something of order 10^{-9} for $k = 12$ or 10^{-6} for $k = 6$. By contrast, the Donaldson algorithm only achieves the $k = 6$ value for H_{MA} when used at the $k = 12$ level and scales as $H_{\text{MA}} \sim \frac{\alpha}{k^2} + \frac{\beta}{k^3}$. This is inline with expectations from the above discussion, and it should be noted that performing the Donaldson computation at $k = 12$ takes roughly two days on a modern desktop machine.

What are the errors in the results that are obtained? In terms of how close we are to the optimal metric for a given k, these can be estimated by a bootstrap method [37]. We compute the metric multiple times and then calculate the standard deviation of the results. In Ref. [37], this was carried out, and it was found that the error was down by a factor of $1/\sqrt{N_p}$ from the expected values. In short, these errors are negligible when compared to those resulting from working at finite k. However, are the finite k approximations that are obtained good enough? This depends entirely on the application that one requires the Ricci-flat metric for! Not many such applications have been pursued yet in the literature (see, for example, Refs. [13,21]) due to the drawbacks that these conventional numerical techniques suffer from. It is to these that we now turn.

The conventional numerical approaches to finding approximate Ricci-flat metrics on CY manifolds that we have been discussing in this section suffer from two qualitative drawbacks:

- For the computations to be practically feasible, symmetries must be imposed upon the CY manifolds of interest. As described earlier, this reduces the number of parameters that need to be considered in our *ansätze* for the Kähler potential. Unfortunately, in many applications, one would like to consider more generic CY manifolds.

- All of the computations that have been carried out in the literature with conventional numerical techniques are performed at a single point in moduli space. This is a severe limitation for most applications. Often, one wishes to know this metric so that it is possible to compute those parts of the effective theory of a string theory compactification that depend on its exact form. We want those potential and kinetic terms as a function of the moduli in almost all instances.

It is the last of these issues in particular that has lead to very little progress being made with the numerical metrics that we have managed to compute in the past. Fortunately, both of these drawbacks can be addressed by the use of ML techniques. For example, there is work computing Ricci-flat metrics as a function of complex structure moduli [5, 32]. In addition, in almost all of the ML literature in this subject, there is no need to impose any symmetry constraint on the CY manifold. One can also pursue more novel approaches in the ML context, for example describing g directly with a neural network rather than using a Kähler potential *ansatz* such as (5.14). This has allowed us to investigate non-Kähler string compactifications as well [5]. It is to these ML-based approaches to the problem of finding metrics on compactifications manifolds that we turn in the following section.

5.3 Lectures 4 and 5: Ricci-Flat Metrics from Neural Networks

By now, it should be evident that the implementations of Donaldson's algorithm [28], as well as the numerical minimization of Headrick and Nassar [37], have practical limitations. These numerical methods rely on an algebraic *ansatz* where the Kähler potential is expanded in a spectral basis (homogeneous polynomials of degree k). This has an inherent scaling problem: the number of monomials N_k grows with k, and the number of parameters of the expansion grows as N_k^2. An additional conundrum is that it is computationally non-trivial to restrict the ambient space polynomials to the CY manifold by quotienting the spectral basis by the ideal of the defining polynomial.[3]

At special points in the CY moduli space, where the polynomial basis is constrained by discrete symmetry groups, this scaling problem is

[3]The restrictions requires a Groebner basis analysis, which is also computationally costly for moderate k.

circumvented by restricting the expansion of the Kähler potential to invariant elements of the basis. Thus, whenever we are interested in the Ricci-flat metric at such special points, the methods at hand suffice. This is true for the Fermat and Dwork quintics, and this is the main reason why these CY manifolds have been favoured in the literature on numerical CY metrics.

However, this means that traditional numerical methods often fail at generic points in moduli space (for a counterexample, see Ref. [12]). Similarly, the methods run into trouble for the vast majority of CY manifolds that are not defined as one hypersurface in one projective ambient space. The CICY manifolds, classified by Candelas *et al.* [20], and the CY hypersurfaces in the toric ambient spaces, classified by Kreuzer and Skarke [44], provide example geometries where, for $k \geq 2$, the dimension N_k is prohibitively large.

This is the main motivation to employ ML methods to approximate the Ricci-flat metric g_{CY} on CY manifolds. After all, the metric we seek to approximate is determined by a function, the Kähler potential, and the universal approximation theorems state that a sufficiently elaborate neural network can approximate any function. In these lectures, we review the various approaches that have recently been employed for such experiments [4, 5, 7, 26, 32, 43, 46, 47]. We also comment on how such methods can improve on the scope and accuracy for CY metric approximation. One aim is to allow you to use these tools for your own research, so the lecture will include practical experiments and refer to open-source code that you may download and experiment with. To date, there are three packages available for ML of CY metrics, namely:

`cymetric`: https://github.com/pythoncymetric/cymetric
`MLgeometry`: https://github.com/yidiq7/MLGeometry
`cyjax`: https://github.com/ml4physics/cyjax

The first of these packages, `cymetric`, currently has the broadest range of applicability, and will therefore be the main focus of these lectures. For the concrete example studied in the following section, any of these packages will do.

5.3.1 *Machine learning on the quintic*

To keep our discussion concrete, let ut set up the problem. We want to find a Ricci flat metric on a Calabi–Yau 3-fold, which we will assume is constructed either as a CICY [20] or as a hypersurface in a toric ambient space from Kreuzer and Skarke's list [44]. For definiteness, we may take this manifold to be the quintic, X, defined as a hypersurface in $\mathbb{P}^4$, as the zero locus of a quintic polynomial $p(Z) = 0$. To be even more specific, we can take the polynomial to be

$$p = \sum_{a=0}^{4} (Z^a)^5,$$

where Z^a are the homogeneous coordinates of $\mathbb{P}^4$. This would define the Fermat quintic, arguably the simplest CY 3-fold we can think of. Any other quintic polynomial would do and would provide another space in the family of quintics, which is related to the Fermat quintic by a complex structure deformation.[4]

As we know from the CY theorem, finding the Ricci-flat metric (in a given Kähler class) is equivalent to solving the Monge–Ampère equation

$$J \wedge J \wedge J = \kappa \Omega \wedge \overline{\Omega},$$

where κ is a constant.[5] We want to solve this equation for J, which defines the CY metric. To our aid, we have Ω, which can be computed, patch by patch, as a residue in the ambient space; this was explained in the first lecture of this series; see Section 5.1. We also recall that J is related to any known representative in the specified cohomology class, such as the FS form J_{FS}, by

$$J = J_{\text{FS}} + \partial\bar{\partial}\phi$$

for a function ϕ. This is helpful since FS Kähler forms may be constructed by restriction from the corresponding ambient space forms, which, as explained in Section 5.1, we know how to compute.

Thus, our ML implementation should be such that we can compute Ω and J_{FS} in the points we sample on the CY 3-fold. We will then feed this information to a neural network that should provide an approximation of the Ricci-flat metric g or the related Kähler potential K. We will train the network using the constraints we have for the problem, such as the Monge–Ampère equation, which we encode as loss functions to be

---

[4]We could also work in other dimensions, say with a quartic CY 2-fold. As a rule of thumb, reducing the dimension of the CY manifold will speed up ML experiments.

[5]This means that κ is constant with respect to the CY coordinates; it will in general depend on the Kähler and complex structure moduli.

minimized during training. Finally, we should check the accuracy of the approximation, for example by computing the Ricci tensor of the predicted metric. By encoding these checks as error measures, we can trace their evolution during training of the network. We want to do this in a systematic and modular fashion, so that our implementation can be readily adapted to different points in moduli space and eventually to different CY manifolds. A schematic view of the implementation is presented in Figure 5.5.

We see that this ML implementation follows the traditional methods to a large degree. First, the same point sampling routines described above for the quintic may be employed.[6] For concreteness, we use the homogeneous sampling method introduced in Section 5.2.2.3, which gives a sample of points that are randomly distributed with respect to the FS measure

$$d\,\mathrm{Vol}_{\mathrm{FS}} = J_{\mathrm{FS}} \wedge J_{\mathrm{FS}} \wedge J_{\mathrm{FS}}.$$

Second, in order to compare performance of different methods, we use standard error measures known as the σ and Ricci measures. The σ measure, first introduced by Donaldson, is obtained by integrating the Monge–Ampère equation over the CY:

$$\sigma = \frac{1}{\mathrm{Vol}_\Omega} \int_X \left| 1 - \kappa \frac{\Omega \wedge \overline{\Omega}}{(J)^n} \right|, \tag{5.24}$$

where $\mathrm{Vol}_\Omega = \int d\mathrm{Vol}_{\mathrm{CY}} = \int \Omega \wedge \overline{\Omega}$ is the volume of the manifold. The Ricci measure is obtained by integrating the Ricci scalar

$$\mathscr{R} = \frac{\mathrm{Vol}_{\mathrm{CY}}^{1/3}}{\mathrm{Vol}_\Omega} \int_X |R|. \tag{5.25}$$

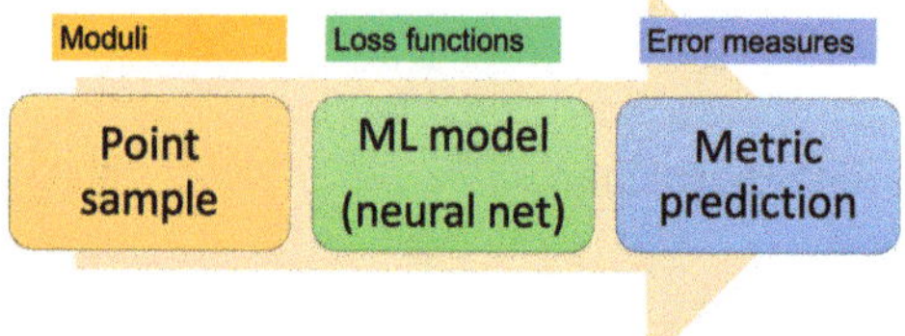

Fig. 5.5. Any ML implementation of CY metric predictions first requires a point sampling routine for the complex structure and Kähler moduli specified by the user. Second, a neural network should be trained on the point sample to satisfy the constraints on the metric, which are encoded as loss functions. Finally, the accuracy should be tested using standard error measures.

[6]Beyond the quintic, new routines will be needed, as we will briefly discuss in what follows.

These measures are closely related to the Monge–Ampère and Ricci functionals introduced in (5.20) and (5.21) and will be evaluated numerically as weighted sums, as discussed in Section 5.4. In general, the integral of a function f is approximated by[7]

$$\int_X \mathrm{dVol}_{\mathrm{CY}} f = \int_X \frac{\mathrm{dVol}_{\mathrm{CY}}}{\mathrm{d}A} \mathrm{d}A\, f = \frac{1}{N} \sum_i m_i f|_{p_i} \quad \text{with} \quad m_i = \frac{\mathrm{dVol}_{\mathrm{CY}}}{\mathrm{d}A}|_{p_i},$$

where the weights/masses m_i is the quotient between the CY measure and the FS measure. These weights should be computed during sampling, so that they can be used whenever an integration is to be computed.

Our main task then is to devise an ML model that predicts the CY metric. This entails specifying what the model should predict: should this be the metric g, the Kähler potential K, or the Hermitian matrix H that defines the algebraic CY metrics of Donaldson's algorithm? All of these choices have been implemented in the literature, and their respective advantages will be discussed to some extent here. Given our choice of prediction, we then need to encode any remaining mathematical constraints as loss functions that guide the training of our network. Finally, we should determine the architecture of our network (the type of layers, activation functions, etc.) and which training algorithm we should use.

To keep this discussion within the space limit of this book, we now focus on the implementation of the `cymetric` package [46, 47], which is constructed to directly predict the Ricci flat metric g. As such, this package includes and extends some of the implementations presented in Ref. [5]. After this discussion, we will comment on alternative neural net designs that are either provided by other packages or could be developed in the future.

5.3.1.1 *Approximate CY metrics from the* `cymetric` *package*

Before describing the implementation, let us take a quick peak under the hood of the `cymetric` package. This will make our discussion more grounded and hopefully give a feeling for how available ML libraries are used in this type of research.

[7] The statistical error with computing integrals as such a weighted sum is roughly $1/\sqrt{N}$, where N is the number of points in the sample, as long as the weights/masses are first-order numbers [25] (this fails in highly curved regions; see Ref. [10] for a discussion).

The `cymetric` package is designed for CY metric prediction on generic CY manifolds at any specified point in their moduli space (e.g., there is no need to restrict to points where the manifold is invariant under large symmetry groups). It decomposes into point sampling routines and ML models.[8] The point generators use functionality of `NumPy` [36] and `SciPy` [56], combined with routines from SageMath [52] and Mathematica. The ML models are implemented and optimized with TensorFlow [1]. This part of the package is structured into python classes that inherit functionality from TensorFlow superclasses and various helper functions that are used to call, compile, and train models. The models are designed to be general, but care has been taken to write an efficient code, so that ML experiments on non-trivial manifolds can be run without the need for large computing resources.[9]

Setting up the ML models. Let us assume that we have generated a sample of points on the Fermat quintic. This provides the input to the neural network that we construct to predict the CY metric. In `cymetric`, the coordinates of each sampled point is saved as a tuple of twice the homogeneous dimension of the ambient space,[10] i.e., $2 * 5$ for $\mathbb{P}^4$. This takes into account that neural networks operate better with real numbers, rather than complex; an input layer with 10 nodes can then read in the real and imaginary parts of the coordinates.

The output layer of the network should give the metric evaluated at the point specified by the coordinates we fed into the input layer. This is naively a 6×6-dimensional matrix; however, the Hermitian structure allows us to cut down the number of components to only predict to the nine entries of a 3×3-dimensional, positive definite, Hermitian matrix (cf. Eq. (5.4)). Thus, the output layer of our neural net should be 9-dimensional.

[8]The package is written in Python and Mathematica [42], and we will predominantly refer to the Python version.

[9]For example, the package uses the so-called function decorators of TensorFlow in order to speed up computations.

[10]This tuple thus contains a redundant description of the three local complex coordinates of the CY; one reason for this design is that it allows to easily identify the coordinate patch the point belongs to, and this information helps in constraining how the metric transforms between coordinate patches.

In fact, the `cymetric` package allows the user to choose whether the NN should predict the full CY metric, as described in the previous paragraph, or a correction to the FS metric that is already known. The latter is likely more efficient, as a randomly initiated neural net may very well predict something very far from the CY metric (e.g., a negative definite matrix). The package provides a number of metric *ansätze* of the type $g = g_{\text{FS}} + \cdots$, as presented in Table 5.1. The last of these, the ϕ-model, differs in that it predicts the correction to the FS Kähler potential, not the FS metric; as such, the output layer for a neural net with this *ansatz* should be one-dimensional.

Of course, specifying the input and output layers does not define a trainable neural network. We must define hidden layers, with parameters that can be adjusted during training. Observations in the literature show that simple, shallow (2–3 hidden layers), fully connected, feed-forward networks trained by stochastic gradient descent work remarkably well. Setting up the neural net in `cymetric` is very simple and uses the underlying TensorFlow/Keras functionalities. For example, by writing

```
nn = tf.keras.Sequential()
    nn.add(tfk.Input(shape=(nIn)))
    for i in range(2):
            nn.add(tfk.layers.Dense(nHidden, activation=act))
    nn.add(tfk.layers.Dense(nOut, use_bias=False))
```

we set up a dense feed-forward neural net that consists of an input layer with `nIn` nodes, two hidden layers with `nHidden` nodes and a nonlinear activation function given by act, and an output layer with `nOut` nodes. (see the `cymetric` package's GitHub page for the full example notebook). Thus, once the user has specified hyperparameters such as the width, depth, and activation function, the ML libraries will do the rest of the

Table 5.1. Different ways to encode the Ricci-flat metric.

Name	Ansatz
Free	$g = g_{\text{NN}}$
Additive	$g = g_{\text{FS}} + g_{\text{NN}}$
Multiplicative, element-wise	$g = g_{\text{FS}} + g_{\text{FS}} \odot g_{\text{NN}}$
Multiplicative, matrix	$g = g_{\text{FS}} + g_{\text{FS}} \cdot g_{\text{NN}}$
ϕ-model	$g = g_{\text{FS}} + \partial\bar{\partial}\phi$

initialization. This makes experimenting with different choices for these hyperparameters straightforward, and the reader is encouraged to do so using the example notebooks.[11] As already mentioned, simple NN designs give good results, but any design is possible; thus, there is room to experiment by adding other types of layers when specifying the neural network.

Since the CY metric is unknown, the network is trained with the objective of minimizing custom loss functions, rather than standard loss functions that encode the difference between the prediction and a known target. This is accomplished using backpropagation and stochastic gradient descent, just as described in Fabian Ruehles's first lecture. In short, this training method computes gradients of the loss functions with respect to the network parameters using the chain rule to propagate backwards from the last to the fist layer of the network. The NN parameters are then updated in the direction of steepest gradient descent. This method has elaborations to avoid getting stuck in local minima, e.g., the parameter updates are done using smaller subsets (batches) of the input data (i.e., points on the CY), rather than the full dataset. All of this functionality is provided by TensorFlow and Keras. In particular, derivatives of loss functions are efficiently computed using the so-called gradient tapes. The ML libraries also provide several different optimizers that allow the user to specify elaborations of the backpropagation algorithm.

Once the neural net is trained, we can call it with the coordinates of any point of the CY to get the approximation of the Ricci-flat metric evaluated at this point. Thus, the trained network is the (numerical approximation of the) CY metric. But how do we set up the loss function for this training? This is the topic of the following paragraph.

Training with custom loss functions. Once we have set up the network, we need to build an ML model that we can train. To do so, we pick the metric *ansatz* of our choice from Table 5.1. At the initialization[12] of the model, we must also specify which of the custom loss functions the training should be governed by.

---

[11]When it comes to activation functions, it has been observed that GELU gives better results than ReLU (possibly related to the problem of dying nodes of the latter), but there is room for experimentation here as well.

[12]By default, the parameters of the network are initialized to random numbers drawn from a Gaussian distribution (default is $\mathcal{N}(0,1)$ for the free network and $\mathcal{N}(0,0.01)$ for the other metric *ansätze*).

As mentioned above, the custom loss functions encode all the mathematical constraints that the CY metric should satisfy. Here, we have the Monge–Ampère equation and the Ricci-flatness constraint. We also need to ensure that the predicted metric is Kähler, $dJ = 0$, and stays in the Kähler class of J_{FS}. For some metric *ansätze* like the free and additive ones, it is also necessary to enforce more basic constraints such as a globally defined metric, which transforms correctly on patch overlaps in the atlas that covers the manifold.

In the `cymetric` package, these constraints are encoded in five custom loss functions $\mathscr{L}_i$ that combine to

$$\mathscr{L} = \alpha_1 \mathscr{L}_{\mathrm{MA}} + \alpha_2 \mathscr{L}_{\mathrm{dJ}} + \alpha_3 \mathscr{L}_{\mathrm{transition}} + \alpha_4 \mathscr{L}_{\mathrm{Ricci}} + \alpha_5 \mathscr{L}_{\mathrm{Kclass}}.$$

Here, α_i are tuneable hyperparameters of the model (default value $\alpha_i = 1.0$) that are set by the user when calling the model. The minimization of these loss functions enforce that the predicted metric satisfies the MA equation and is Ricci-flat, Kähler, globally defined, and in the prescribed Kähler class. The first two loss function are, by the CY theorem, encoding the same constraint, so it should be enough to enforce only one of these during training.

In more detail, the first four loss contributions are given by

- Monge–Ampère (MA) loss $\mathscr{L}_{\mathrm{MA}}$

$$\mathscr{L}_{\mathrm{MA}} = \left\| 1 - \frac{1}{\kappa} \frac{\det g_{\mathrm{pred}}}{\Omega \wedge \bar{\Omega}} \right\|_n ,$$

- Ricci loss $\mathscr{L}_{\mathrm{Ricci}}$[13]

$$\mathscr{L}_{\mathrm{Ricci}} = \|R\|_n = \|\partial\bar{\partial} \ln \det g_{\mathrm{pred}}\|_n ,$$

- Kähler loss $\mathscr{L}_{\mathrm{dJ}}$

$$\mathscr{L}_{\mathrm{dJ}} = \sum_{ijk} \|\mathrm{Re}\, c_{ijk}\|_n + \|\mathrm{Im}\, c_{ijk}\|_n ,$$

where $c_{ijk} = g_{i\bar{j},k} - g_{k\bar{j},i}$ and $g_{i\bar{j},k} = \partial_k g_{i\bar{j}}$.

[13]This is strictly only a good loss function for Kähler metrics, where the Ricci tensor simplifies to $R_{i\bar{j}} = \partial_i \bar{\partial}_{\bar{j}} \log \det(g)$.

- Transition loss $\mathscr{L}_{\text{transition}}$

$$\mathscr{L}_{\text{transition}} = \frac{1}{d} \sum_{\mathcal{U},\mathcal{V}} \left\| g^{\mathcal{V}}_{\text{pred}} - T_{\mathcal{U}\mathcal{V}} \cdot g^{\mathcal{U}}_{\text{pred}} \cdot (T_{\mathcal{U}\mathcal{V}})^{\dagger} \right\|_n ,$$

  where d is the number of patch transitions and the transition matrices $(T_{\mathcal{U}\mathcal{V}})^{\nu}_{\mu} = \partial v^{\nu} / \partial u^{\mu}$ are the Jacobians of the coordinate transform between the coordinates u in $\mathcal{U}$ and v in $\mathcal{V}$.

These losses are computed over mini-batches containing N_B points, and as indicated by the subscript n, the loss functions can be computed with any L_n norm[14]:

$$\|x\|_n = \left(\sum_{i=1}^{N_B} x_i^n \right)^{1/n} .$$

For all these loss functions, the derivatives with respect to the input coordinates are computed with TensorFlow's automatic differentiation but with the twist that gradient tapes are applied to the input, not network parameters. There is a computational cost associated with this, particularly for the Ricci loss, which requires evaluating two derivatives. Therefore, this loss function, which is in principle encoding the same constraint as the MA loss on CY geometries, is disabled by default in the package.

There is a final component in the loss which enforces that the predicted metric stays in the correct Kähler class. Since this loss function is of little interest for experiments on the quintic or any other CY manifolds with $h^{(1,1)} = 1$, we defer the description of this loss function to the following section.

With these loss functions defined, we can initialize and train the ML model using the full functionality of TensorFlow. This is simply done by first initializing the chosen model with the neural net, information about CY geometry, and loss function hyperparameters as inputs. For example,

```
fmodel = AddFSModel(nn, BASIS, alpha=[1., 1., 1., 1., 1.])
```

[14]The default in the package is $n = 1$ for all losses but $\mathscr{L}_{\text{dJ}}$, which has $n = 2$. One should choose the L_n norm wisely, as it affects the training of the network. With $n = 1$, all points in the batch are treated equally, and the network strives to minimize the average loss. With higher n, the network prioritizes outliers, i.e., points with the largest loss function contributions.

would initialize an ML model corresponding to the additive metric *ansatz*, with default values of the loss hyperparameters α, which has access to the information of the CY contained in `BASIS`.[15]

Second, we compile the model and then train it using the custom loss function on the sampled points. This is done using the `compile` and `fit` commands of TensorFlow/Keras. To simplify the computation of the various components of the loss, we call these functions using the `train-model` helper function of the `cymetric` package, i.e.,

```
fmodel, training_history = train_model(fmodel, data,
    optimizer=opt, epochs=N, batch_sizes=[64, 50000],
    verbose=1, custom_metrics=cmetrics, callbacks=cb_list)
```

This command starts a training loop that will run over N epochs, where each epoch consists of a first training loop over all mini-batches (containing 64 points) and a second training loop over the larger batches (containing 50,000 points). This two-step training will, as alluded to above, first update the network parameters to minimize the MA, Kähler, and transition losses.[16] For the quintic, the second training step (over the large batches) is unnecessary as its objective, preservation of the Kähler class, is automatic.[17] We will describe this loss function in more detail in the following section.

As evident from the code snippet above, when we train the model, the compile and train commands must also specify callbacks and custom metrics, which allows us to track the training dynamics. The package's custom metrics consist of the loss functions (including the Ricci loss), so that we can check that these decrease during the training process. The callbacks are measures of the accuracy of the metric prediction that we will evaluate on test data. These include the σ and Ricci measures in Eqs. (5.24) and (5.25), which are benchmark measures [2,24] used to compare the accuracy of any numerical approximation of the CY metric, including the algebraic

---

[15]This data container is created by the point generators of `cymetric` and is described in detail in the package's documentation. It holds information such as the constant κ, the defining polynomial for the CY, etc.

[16]Recall that the Ricci scalar is disabled by default.

[17]To avoid this unnecessary step, the `compile` and `fit` commands may be used directly. Alternatively, the user may modify the `train-model` function by removing the second training step. For the following experiments, the second training step only adds 1 s per epoch, in comparison to the 30 s needed for the first step. Thus, the time saved is not significant, but for longer training on more complicated quintics, it might be.

metrics that result from Donaldson's algorithm or Headrick and Nassar's functional methods. In addition to these standard measures, there are also callbacks that check the constancy of the volume and that the metric prediction is Kähler and globally defined. These callbacks are needed to monitor the ML implementation of the `cymetric` package, where these are soft constraints, that are imposed through the loss function.

To conclude, we have described what we need to do to machine learn Ricci-flat CY metrics. To summarize, we must:

(1) generate points,
(2) create an NN,
(3) choose an ML model (i.e., metric *ansatz*),
(4) train the model,
(5) check accuracy.

Let's gauge the computational resources such experiments takes. On a standard laptop,[18] generating some 100,000 points takes from a few minutes to half an hour for CICY manifolds with moderate number of Kähler moduli and a bit longer for CYs in toric ambient spaces. Training the model for 50 epochs takes half an hour on simple quintics and in the order of hours for more elaborate CYs; as one would expect, the simpler (more symmetric) the geometry, the faster the training. Using a machine with a GPU leads to faster execution.

Simple experiments. We are now ready to perform experiments using the `cymetric` package. Let's start simple and run one of the ML models on the Fermat quintic. We will pick the ϕ model with default loss function weights. After generating 100,000 points on the manifold (which we split 9:1 into a training and validation set, where the latter is hidden during training), we set up a fully connected NN with three hidden layers of width 64. We select GELU activation functions, the Adam optimizer, and train for 100 epochs (about 1 hour). The results of the training are given in Figure 5.6. We see that the MA loss decreases, during training, by a factor 20. This is true for both the training and validation sets. We see that the Kähler loss is vanishing throughout the training and that the transition loss functions decrease by a factor of 10 (both evaluated on the validation set). Finally, we plot how the Ricci and sigma measures decrease

[18]To be more precise, on a 2016 MacBookPro that has seen better days.

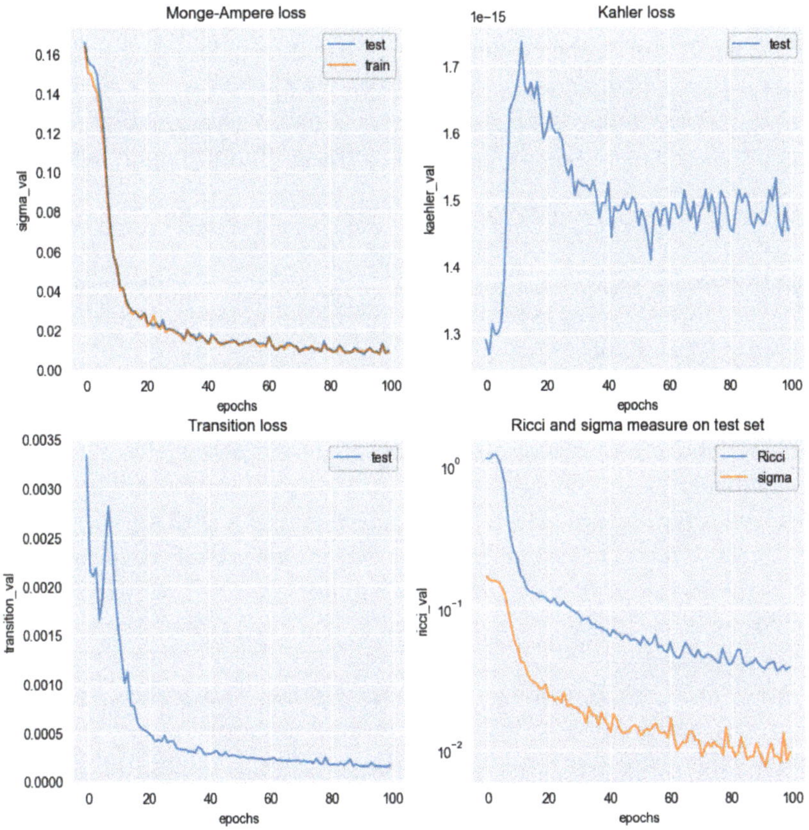

Fig. 5.6. A simple experiment using the ϕ network on the Fermat quintic. Training clearly reduces the MA and transition loss and keeps the Kahler loss negligible. The error measures, computed on points not seen during training, also decrease.

during training, reaching a level of 0.05 and 0.01, respectively. These training results clearly show that learning is taking place.

Note that we have looked at a very simple network, with the most standard set up for the training. The models may be optimized in many different ways — indeed, the non-zero slopes of the losses and error measures even indicate that simply training longer will improve the results. The results are nevertheless proof that the ML methods work (and that work towards better precision is motivated, as we will discuss in what follows).

What is more, these ML methods have a wide range of applicability. We can easily run the experiment that we just did, but for a quintic defined by a different polynomial. As an example, we can construct a completely generic quintic. Following Ref. [13], we pick random, complex coefficients a_i of norm $|a_i| \in (0,1)$ for the 126 homogeneous quintic monomials

$$p = \sum a_i m(Z) = a_0(Z^0)^5 + a_1(Z^0)^4 Z^1 + \cdots + a_{126}(Z^4)^5, \tag{5.26}$$

This breaks all discrete symmetries of the quintic, and thus using traditional numerical methods to compute the Ricci-flat metric would require extensive computational power.

We again generate 100,000 points on this non-symmetric quintic. Running the exact same experiment as for the Fermat quintic takes again of the order of hours. The results of the training are presented in Figure 5.7. Qualitatively, these look very similar to the Fermat results: the loss functions and error measures decrease during training. Quantitatively, the values for the MA and transition losses are larger than in the Fermat case, and the loss decrease is not as big (factor of 5 rather than 20 for MA and transition loss; the Kahler loss is still vanishing). The gain on the error measures is a factor of 2–3.

While it is hard to draw conclusions from these experiments without more systematic studies,[19] one may observe that the "non-symmetric" model generalizes less well to the test set. This difference in generalizability is likely attributed to symmetry. Recall that the Fermat quintic is a highly symmetric manifold. Its defining polynomial is invariant under discrete $\mathbb{Z}_5$ rotations, permutations, and complex conjugation of the homogeneous coordinates.[20] This symmetry must be learned to some extent by the network, and this will help it to generalize to new points on the manifold. On the other hand, the quintic defined by (5.26) breaks these symmetries. As such, the network must approximate a more complicated metric and will likely be more complicated and possibly more prone to fitting to non-generalizable features of the training data. Interestingly, despite this difference in training results, the final value

[19]Recall the large amount of randomness inherent in these experiments; we sample random points, we train with stochastic gradient descent, etc., and so fluctuations in performance are expected. It is good practice to run a number of different experiments and draw conclusions based on average performance.

[20]Taking the projective rescaling into account, the full symmetry group of the Fermat quintic is $\mathbb{Z}_5^4 \rtimes (S^5 \times \mathbb{Z}_2)$ [13].

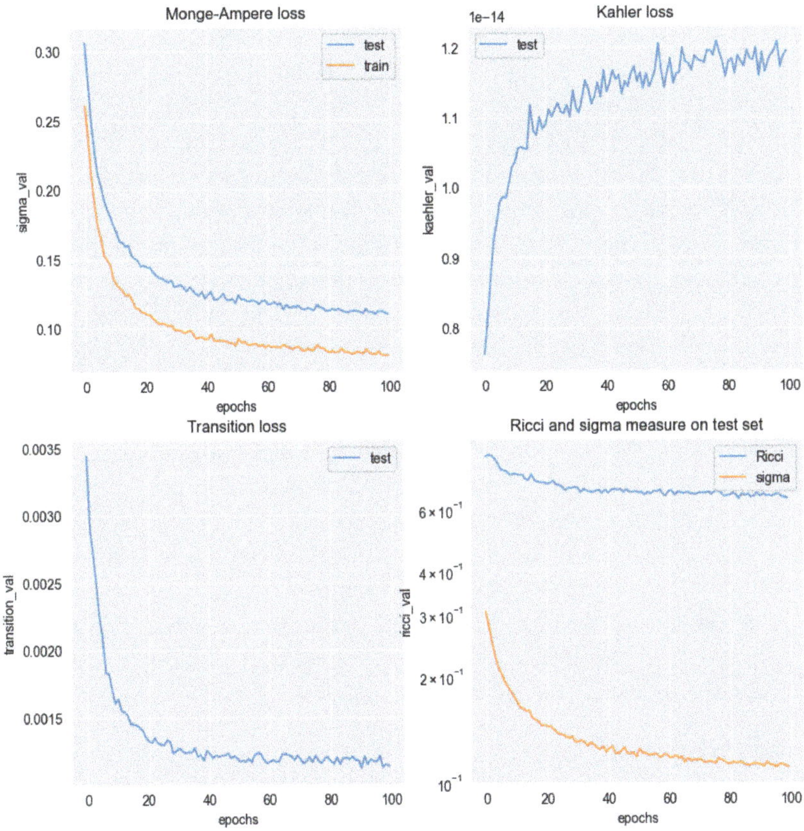

Fig. 5.7. Training the ϕ network on a quintic at a non-symmetric point in moduli space. Training takes slightly longer than on the Fermat quintic (notice the non-vanishing slope in the MA loss), and the results do not generalize as well to the test set. However, qualitatively, the plots are very similar; clearly, learning is taking place also on this manifold.

for the Ricci and sigma measures are comparable in the two experiments. This indicates that the ML performance is similar on the two quintics.

The main point of performing this second experiment on the non-symmetric quintic is that we are no longer limited to working with highly symmetric CY manifolds. Indeed, the direct ML approach also works far beyond the regime accessible for algebraic metrics. With ML, we can look at any example from the CY databases, and this is very exciting for string theory (and mathematics).

5.3.2 *Machine learning experiments beyond the quintic*

Most CY manifolds in the CICY and Kreuzer–Skarke databases differ from the quintic in one of the following respects: they may have $h^{(1,1)} > 1$, be defined by more than one homogeneous polynomial, and/or be defined as a hypersurface in a toric ambient space rather than a projective one. This calls for new point generators and a new training loop. Such point-generating algorithms have been developed for CICY manifolds [12] and CY manifolds in toric ambient spaces [46, 47]. They are direct generalizations of the homogeneous sampling technique described for the Quintics in Section 5.2.2.3 and are included in the `cymetric` package's point generator routines. Describing how this works in detail would require another lecture. So, instead, we will refer the reader to Ref. [47], where this topic is discussed in detail. We will instead focus on the training loop modifications that are needed for more elaborate CY manifolds.

For manifolds with more than one Kähler class, there is a risk that training will push the predicted metric out of the initial Kähler class. This is problematic: the constant κ is computed in this Kähler class, and the MA equation loses its meaning if the network jumps between classes. To ensure this does not happen, a new loss function must be added. In the `cymetric` package, the following loss is defined:

- Kähler class loss $\mathcal{L}_{\text{Kclass}}$

$$\mathcal{L}_{\text{Kclass}} = \frac{1}{h^{1,1}(X)} \sum_{\alpha=1}^{h^{1,1}(X)} \left\| \mu_t(\mathcal{O}_X(e_\alpha)) - \int_X J_{\text{pr}}^2 \wedge F_{FS,\alpha} \right\|_n . \quad (5.27)$$

  This function computes certain topological quantities, the slopes μ_t, for a basis of line bundles $\mathcal{O}_X(e_\alpha)$, $\alpha = 1, \ldots, h^{1,1}(X)$. These line bundles are specified by the integers $(e_\alpha)^\beta = \delta_\alpha^\beta$, and their slopes $\mu_t(\mathcal{O}_X(e_\alpha))$ are topological quantities which depend on the choice of Kähler class through the Kähler moduli t^α. They can be computed using the intersection numbers of the manifold.[21] Being topological, these quantities should be the same for all metric choices. The slope may also be computed as the integral after the minus sign in (5.27), where J_{pr} is the predicted Kähler form and

$$F_{FS,\alpha} = -iJ_\alpha/(2\pi)$$

[21] The intersection numbers are topological invariants of the CY manifold; they are known for CICY and KS CY manifolds.

are the line bundle field strengths computed using the FS Kähler form.[22] This integral is evaluated numerically as a weighted sum over the sampled points. This is not sensible to evaluate on a mini-batch of around 100 points. Instead, it should be evaluated on a larger batch.

With this addition to the loss function, we can learn Ricci-flat metrics on CY manifolds with multiple Kähler moduli. Given a point sample, we set up a neural net for our preferred metric *ansatz*, as described above. Subsequently, we compile and train the model on the sampled points using the same `train-model` command. This starts a training loop that runs over N epochs using the optimizer, custom metrics, and callbacks we choose to specify. During training, each epoch consists of a first training loop over all mini-batches (containing 64 points) and a second training loop over the larger batches (containing 50,000 points). This two-step training will first update the network parameters to minimize the MA, Kähler, and transition losses, and secondly, the Kclass loss.[23] The optimizer (with specified learning rate and momentum) is shared between the two optimization steps, and gradients are accumulated in the training.

The double training loop, with its numerical slope computation, makes training slightly slower than using TensorFlow's standard command. However, there is no way around this: we must ensure that the Kähler class is preserved in order to solve the MA equation, and the Kclass loss of the `cymetric` package is, at the time of writing these notes, the only way to ascertain whether this will be true on CY manifolds with $h^{(1,1} > 1$.

5.3.3 *Outlook and open problems*

In these lectures, we have described numerical methods that can be used to approximate Ricci-flat CY metrics. For the ML implementations, we have centred our discussion around one particular tool, the `cymetric` package. With this package, we can compute approximate CY metrics on any CY manifold in the CICY and KS databases at any given point in their moduli space. This package encodes the metric in a neural network, which is trained on a random sample of points on the CY manifold. After training,

[22]For CY manifolds with more than one Kähler modulus, the FS Kähler form is given by $J_{FS} = \sum t^{\alpha} J_{\alpha}$, where t^{α} are the Kähler moduli.

[23]Recall that the Ricci scalar is disabled by default.

the metric can be computed at any given point on the manifold by simply calling the trained ML model.

As already mentioned, there are two other open-source packages, `MLgeometry` [26] and `cyjax` [32], that also may be used to predict CY metrics. These packages are similar to `cymetric` in that they use fully connected neural networks, trained by stochastic gradient descent. They differ in some architectural choices for the neural nets, the ML libraries they rely on, and in how points are sampled and presented to the ML models. One important difference is that they encode more of the CY constraints in the design of the neural net. Rather than predicting the metric, these networks predict the Kähler potential K or Donaldson's H matrix. Second, global consistency is ensured by expanding K in homogeneous polynomials of degree k, similarly to what is done for the numerical methods using algebraic metrics. With this alternative design, there is no need to encode global consistency or Kählerity constraints as loss functions; in fact, only the MA loss is used in the training of these networks. However, an analogue of the Kclass loss of `cymetric` would be needed for these packages to work on CY manifolds with $h^{(1,1)} > 1$; this is lacking at the time when these notes where written.

The encoding of constraints in the network is advantageous when it comes to the accuracy of the prediction, a topic discussed in, e.g., Ref. [26]. It has also made it possible to explore the complex structure moduli dependence of the CY metric approximation [5, 32]. However, encoding more information in the network leads to designs that are harder to generalize to new settings. Indeed, these models are, to date, tailor-made for CY spaces with a single Kähler class. Thus, they require some adaption to work on general CY manifolds. One disadvantage in such generalizations would then be the curse of dimensionality associated with the dimension of the polynomial basis, which will be the same as in the "pre-ML" numerical methods. An interesting middle ground, dubbed the "spectral network", was recently proposed in Ref. [14]. Here, the general implementation of the `cymetric` package is combined with an expansion of the metric in a basis of homogeneous polynomials. This is shown to improve the accuracy of metric predictions, particularly in nearly singular quintic manifolds, in agreement with the findings of Ref. [26].

As we conclude these lectures, it is reasonable to ask how well these ML methods perform relative to their more conventional counterparts? Such comparisons have been made in the literature, usually by comparing the ML methods to the Donaldson algorithm (see, e.g., Refs. [5, 26]).

However, as was described in Section 5.2.3, standard functional minimization already achieves essentially the optimal metric, which is closest to Ricci-flat for the given *ansatz*, up to errors that are negligible compared to working at finite k. Thus, one might think that at a given k, ML methods cannot do any better. This has been shown to be incorrect for two reasons. Firstly, as already explained, ML methods can compute these metrics without using the symmetry assumptions that are required in the conventional approach. We can thus perform computations in a few hours or even a few minutes in optimized settings, at points in moduli space that simply do not finish with conventional techniques. Secondly, some results which have been obtained in the ML contexts have been computed as a function of complex structure moduli [5,32]. This too has never been achieved with more conventional approaches. Thirdly, with alternative network architectures, training of ML models can speed up significantly; see Ref. [38] for an exciting new proposal that, in a very short time, provides highly accurate approximations for the Ricci-flat metric on the training data from the Fermat quintic. However, the models in this reference do not generalize well; in the current form, they do not provide accurate predictions on test data.

Hopefully, these lectures have demonstrated that numerical and ML methods are useful in exploring CY geometries. Undoubtedly, there is also scope to improve these methods further. Indeed, it is important to stress that this is a field of research where there are many open questions. In particular, the field of ML and CY geometry is still in its infancy and breakthroughs regarding efficiency, accuracy, and scope are likely to emerge. One very interesting open question is how to approximate moduli-dependent CY metrics beyond the partial explorations of the quintic's complex structure moduli space done in Refs. [5,32]. Another interesting application is to use numerical CY metrics to identify regions of high curvature on CY manifolds at different points in their moduli space; see Refs. [14,21] for recent discussions.

Numerical methods also allow us to compute massive spectra of string compactifications, with relevance to, e.g., the Swampland program; see Refs. [9–11, 13]. These numerical methods also allow us to compute masses and couplings related to vector bundles over CY manifolds; see Refs. [2,3,6]. In a different direction, numerical and ML methods have only very rarely been applied outside the CY regime. Other geometries are also of interest in string theory, such as different types of G structure geometries. As demonstrated in Ref. [5], one can make progress in some of these directions using the ML techniques described here. Indeed, with general ML models, one need only adapt the loss functions to encode, instead of the CY constraints, whatever restrictions such metrics should satisfy. In Ref. [5], this allowed the authors to machine learn an SU(3) structure on the quintic, which was shown to exist in Ref. [45].

Finally, returning to the CY case, it is fascinating that the simplest type of neural networks can make progress on such a non-trivial mathematical problem. Does this mean that Ricci-flat CY metrics are, perhaps, simpler than we have thought? Perhaps, by analyzing models trained on different CY geometries, we can hope to detect universal properties of their Ricci-flat metrics? Perhaps, you will contribute to answering this question (hopefully in the positive) in the future?

References

[1] M. Abadi, A. Agarwal, P. Barham, E. Brevdo, Z. Chen, C. Citro, G. S. Corrado, A. Davis, J. Dean, M. Devin, S. Ghemawat, I. Goodfellow, A. Harp, G. Irving, M. Isard, Y. Jia, R. Jozefowicz, L. Kaiser, M. Kudlur, J. Levenberg, D. Mané, R. Monga, S. Moore, D. Murray, C. Olah, M. Schuster, J. Shlens, B. Steiner, I. Sutskever, K. Talwar, P. Tucker, V. Vanhoucke, V. Vasudevan, F. Viégas, O. Vinyals, P. Warden, M. Wattenberg, M. Wicke, Y. Yu, and X. Zheng, "TensorFlow: Large-scale machine learning on heterogeneous systems," 2015. Software available from tensorflow.org.

[2] L. B. Anderson, V. Braun, R. L. Karp, and B. A. Ovrut, "Numerical hermitian Yang-Mills connections and vector bundle stability in heterotic theories," *JHEP* **06**, 107, 2010.

[3] L. B. Anderson, V. Braun, and B. A. Ovrut, "Numerical hermitian Yang-Mills connections and kahler cone substructure," *JHEP* **01**, 014, 2012.

[4] A. Ashmore, L. Calmon, Y.-H. He, and B. A. Ovrut, "Calabi-Yau metrics, energy functionals and machine-learning," *Int. J. Data Sci. Math. Sci.* **1**, no. 1, 49–61, 2023.

[5] L. B. Anderson, M. Gerdes, J. Gray, S. Krippendorf, N. Raghuram, and F. Ruehle, "Moduli-dependent Calabi-Yau and SU(3)-structure metrics from machine learning," *JHEP* **05**, 013, 2021.

[6] A. Ashmore, Y.-H. He, E. Heyes, and B. A. Ovrut, "Numerical spectra of the Laplacian for line bundles on Calabi-Yau hypersurfaces," *JHEP* **07**, 164, 2023.

[7] A. Ashmore, Y.-H. He, and B. A. Ovrut, "Machine learning Calabi–Yau metrics," *Fortsch. Phys.* **68**, no. 9, 2000068, 2020.

[8] L. B. Anderson and M. Karkheiran, "TASI lectures on geometric tools for string compactifications," *PoS* **TASI2017**, 013, 2018.

[9] A. Ashmore and F. Ruehle, "Moduli-dependent KK towers and the swampland distance conjecture on the quintic Calabi-Yau manifold," *Phys. Rev. D* **103**, no. 10, 106028, 2021.

[10] H. Ahmed and F. Ruehle, "Level crossings, attractor points and complex multiplication," *J. High Energ. Phys.* **2023**, 164, 2023.
[11] A. Ashmore, "Eigenvalues and eigenforms on Calabi-Yau threefolds," *J. Geom. Phys.* **195**, 105028, 2024.
[12] V. Braun, T. Brelidze, M. R. Douglas, and B. A. Ovrut, "Calabi-Yau metrics for quotients and complete intersections," *JHEP* **05**, 080, 2008.
[13] V. Braun, T. Brelidze, M. R. Douglas, and B. A. Ovrut, "Eigenvalues and eigenfunctions of the scalar laplace operator on Calabi-Yau manifolds," *JHEP* **07**, 120, 2008.
[14] P. Berglund, G. Butbaia, T. Hübsch, V. Jejjala, D. M. Peña, C. Mishra, and J. Tan, "Machine learned Calabi-Yau metrics and curvature," *Adv. Theor. Math. Phys.* **27**, 1107–1158, 2023.
[15] M. Berger, "Sur les groupes d'holonomie homogènes de variétés à connexion affine et des variétés riemanniennes," *Bull. Soc. Math. Fr.* **83**, 279–330, 1955.
[16] V. Braun, "On Free Quotients of Complete Intersection Calabi-Yau manifolds," *JHEP* **04**, 005, 2011.
[17] P. Candelas, *Lectures on Complex Manifolds*. Singapore: WSPC, 1987.
[18] P. Candelas, "Yukawa couplings between (2,1) forms," *Nucl. Phys. B* **298**, 458, 1988.
[19] P. Candelas and X. C. de la Ossa, "Comments on conifolds," *Nucl. Phys. B* **342**, 246–268, 1990.
[20] P. Candelas, A. M. Dale, C. A. Lutken, and R. Schimmrigk, "Complete intersection Calabi-Yau manifolds," *Nucl. Phys.* **B298**, 493, 1988.
[21] W. Cui and J. Gray, "Numerical metrics, curvature expansions and Calabi-Yau manifolds," *JHEP* **05**, 044, 2020.
[22] P. Candelas, G. T. Horowitz, A. Strominger, and E. Witten, "Vacuum configurations for superstrings," *Nucl. Phys. B* **258**, 46–74, 1985.
[23] D. A. Cox, J. B. Little, and H. K. Schenck, *Toric Varieties*.
[24] M. R. Douglas, R. L. Karp, S. Lukic, and R. Reinbacher, "Numerical solution to the hermitian Yang-Mills equation on the Fermat quintic," *JHEP* **12**, 083, 2007.
[25] M. R. Douglas, R. L. Karp, S. Lukic, and R. Reinbacher, "Numerical Calabi-Yau metrics," *J. Math. Phys.* **49**, 032302, 2008.
[26] M. R. Douglas, S. Lakshminarasimhan, and Y. Qi, "Numerical Calabi-Yau metrics from holomorphic networks," 2020.
[27] S. K. Donaldson, "Scalar curvature and projective embeddings, II," *Q. J. Math.*, **56**, no. 3, 345–356, 2005, doi: 10.1093/qmath/hah044.
[28] S. K. Donaldson, "Some numerical results in complex differential geometry," *Pure Appl. Math. Q.* **5**, no. 2, 571–618, 2009.
[29] M. R. Douglas, "Holomorphic feedforward networks," *Pure Appl. Math. Quart.* **18**, no. 1, 251–268, 2022.
[30] A. Font and S. Theisen, "Introduction to string compactification," *Lecture Notes in Physics*, Vol. 668, pp. 101–181, 2005.
[31] J. Gray, A. S. Haupt, and A. Lukas, "All complete intersection Calabi-Yau four-folds," *JHEP* **07**, 070, 2013.
[32] M. Gerdes and S. Krippendorf, "CYJAX: A package for Calabi-Yau metrics with JAX," *Mach. Learn.: Sci. Technol.* 4, 025031, 2023.

[33] B. Greene, "String theory on Calabi-Yau manifolds," 1997.

[34] M. B. Green, J. H. Schwarz, and E. Witten, *Superstring Theory. Vol. 2: Loop Amplitudes, Anomalies and Phenomenology*, 1988.

[35] J. Gray and J. Wang, "Free quotients of favorable Calabi-Yau manifolds," *JHEP* **07**, 116, 2022.

[36] C. R. Harris, K. J. Millman, S. J. van der Walt, R. Gommers, P. Virtanen, D. Cournapeau, E. Wieser, J. Taylor, S. Berg, N. J. Smith, R. Kern, M. Picus, S. Hoyer, M. H. van Kerkwijk, M. Brett, A. Haldane, J. F. del Río, M. Wiebe, P. Peterson, P. Gérard-Marchant, K. Sheppard, T. Reddy, W. Weckesser, H. Abbasi, C. Gohlke, and T. E. Oliphant, "Array programming with NumPy," *Nature* **585**, no. 7825, 357–362, 2020.

[37] M. Headrick and A. Nassar, "Energy functionals for Calabi-Yau metrics," *Adv. Theor. Math. Phys.* **17**, no. 5, 867–902, 2013.

[38] J. Halverson and F. Ruehle, "Metric flows with neural networks," *Mach. Learn.: Sci. Technol.* **5**, 045020, 2023.

[39] T. Hubsch, *Calabi-Yau Manifolds: A Bestiary for Physicists*. World Scientific, Singapore, 1994.

[40] D. Huybrechts, *Complex Geometry: An Introduction*.

[41] M. Headrick and T. Wiseman, "Numerical Ricci-flat metrics on K3," *Class. Quant. Grav.* **22**, 4931–4960, 2005.

[42] Wolfram Research, Inc. Mathematica, Version 12.3.1. Champaign, IL, 2021.

[43] V. Jejjala, D. K. Mayorga Pena, and C. Mishra, "Neural network approximations for Calabi-Yau metrics," *JHEP* **8**, 105, 2022.

[44] M. Kreuzer and H. Skarke, "Complete classification of reflexive polyhedra in four-dimensions," *Adv. Theor. Math. Phys.* **4**, 1209–1230, 2002.

[45] M. Larfors, A. Lukas, and F. Ruehle, "Calabi-Yau manifolds and SU(3) structure," *JHEP* **01**, 171, 2019.

[46] M. Larfors, A. Lukas, F. Ruehle, and R. Schneider, "Learning size and shape of Calabi-Yau spaces," 2021.

[47] M. Larfors, A. Lukas, F. Ruehle, and R. Schneider, "Numerical metrics for complete intersection and Kreuzer–Skarke Calabi–Yau manifolds," *Mach. Learn. Sci. Tech.* **3**, no. 3, 035014, 2022.

[48] W. H. Press, S. A. Teukolsky, W. T. Vetterling, and B. P. Flannery, *Numerical Recipes in C*, 2nd edn. Cambridge University Press, Cambridge, USA, 1992.

[49] A. Strominger and E. Witten, "New manifolds for superstring compactification," *Commun. Math. Phys.* **101**, 341, 1985.

[50] B. Shiffman and S. Zelditch, "Distribution of zeros of random and quantum chaotic sections of positive line bundles," *Commun. Math. Phys.* **200**, no. 3, 661–683, 1999.

[51] B. Shiffman and S. Zelditch, "Number variance of random zeros on complex manifolds," *Geom. Funct. Anal.* **18**, 1422–1475, 2008.

[52] The Sage Developers, SageMath, the Sage Mathematics Software System (Version 9.4.0), 2021. https://www.sagemath.org.

[53] G. Tian, "On a set of polarized Kähler metrics on algebraic manifolds," *J. Differential Geom.* **32**, 99–130, 1990.

[54] G. Tian, *Canonical Metrics in Kahler Geometry*. Basel, Switzerland: Birkhauser, 2005.

[55] E. Witten, "Symmetry breaking patterns in superstring models," *Nucl. Phys. B* **258**, 75–100, 1985.

[56] W. McKinney, "Data structures for statistical computing in python," in S. van der Walt and J. Millman (eds.), *Proceedings of the 9th Python in Science Conference*, pp. 56–61, 2010.

[57] S.-T. Yau, "On the Ricci curvature of a compact Kähler manifold and the complex Monge-Ampére equation, I," *Commun. Pure Appl. Math.* **31**, no. 3, 339–411, 1978.

[58] S.-T. Yau, "Compact three-dimensional Kahler manifolds with zero Ricci curvature," 1986.

CHAPTER 6

TUTORIALS IN STRING THEORY COMPACTIFICATION

ELLI HEYES, EDWARD HIRST, AND YIDI QI

In complement to Chapter 5, where methods for numerical approximation of Calabi–Yau metrics were introduced, this chapter presents several tutorial problems to demonstrate their practical implementation.

We begin, in Tutorial 1, with a set of pen-and-paper exercises that familiarize the reader with the concepts of manifolds and Ricci curvature. In Tutorial 2, we move on to numerical methods in `python` for solving a differential equation on a complex manifold, providing a more detailed introduction to the computational aspects. Finally, in Tutorial 3, we address the full problem of finding Ricci-flat metrics on Calabi–Yau manifolds using the `cymetric` and `MLGeometry` packages.

6.1 Tutorial 1: Geometry of Calabi–Yau Manifolds

In this tutorial, we review the basic elements of differential geometry required to follow the subsequent tutorials on Calabi–Yau geometry and Ricci-flat metrics. We begin by considering the definitions of real and complex manifolds, and we demonstrate that S^2 is an example of both by constructing suitable atlases. We then examine the Ricci curvature of a hypersurface in complex projective space. By pulling back the Fubini–Study metric, we show that the Ricci form of a degree-d hypersurface differs from the Fubini–Study Kähler form by an exact term, so that its cohomology class is proportional to $(n+1-d)[\omega_{\text{FS}}]$, where $[\omega_{\text{FS}}]$ denotes the Kähler class of the Fubini–Study form. In particular, for hypersurfaces of degree $n+1$, this class vanishes, establishing the Calabi–Yau condition required for the existence of a Ricci-flat Kähler metric.

(1) (a) Using stereographic projection, set up a coordinate atlas on S^2, that is, a collection of open sets and morphisms (U_i, ϕ_i) with $\phi_i : U_i \to \mathbb{R}^2$ satisfying the manifold conditions.
(b) Now treat S^2 as a complex manifold (that is, $\mathbb{CP}^1$), and repeat the exercise, mapping open sets into $\mathbb{C}$. Do this using homogeneous coordinates $(x_0, x_1) \sim \lambda(x_0, x_1)$, and find the transition functions.

(2) Prove that if X is a degree-d hypersurface in $\mathbb{CP}^n$, then the restriction of the Fubini–Study metric to X leads to a Ricci two-form:

$$\mathscr{R}_{\mathrm{FS}}\big|_X = \frac{2(n-d+1)}{i}\,\omega_{\mathrm{FS}} + \text{total derivatives}.$$

Hence, there exists a metric in the same Kähler class as a Ricci-flat one.

Solutions

(1)(a) We take S^2 to be the set of points in $\mathbb{R}^3$ defined by $x_1^2 + x_2^2 + x_3^2 = 1$.

We construct a chart from the open set U_1, defined as the sphere minus the north pole, via stereographic projection. We draw a straight line from the north pole to the plane defined by $x_3 = -1$ and assign to the point on S^2 intercepted by the line the Cartesian coordinates (y_1, y_2) of the corresponding point on the plane. The parametric equation of the line is

$$\vec{r}(t) = (0,0,1) + t(x_1, x_2, x_3 - 1).$$

It intersects the (x_1, x_2)-plane when $1 + t(x_3 - 1) = -1$, which implies $t = \frac{2}{1-x_3}$. Hence, the point on the plane is given by

$$\phi_1(x_1, x_2, x_3) \equiv (y_1, y_2) = \left(\frac{2x_1}{1-x_3}, \frac{2x_2}{1-x_3}\right).$$

We construct a second chart from the open set U_2, defined as the sphere minus the south pole, using the opposite stereographic projection onto the plane $x_3 = +1$. The map is

$$\phi_2(x_1, x_2, x_3) \equiv (z_1, z_2) = \left(\frac{2x_1}{1+x_3}, \frac{2x_2}{1+x_3} \right).$$

The charts cover the entire manifold and overlap for $-1 < x_3 < 1$. To check the smoothness of the transition function $\phi_2 \circ \phi_1^{-1}$, we compute ϕ_1^{-1}. First,

$$y_1^2 + y_2^2 = \frac{4(x_1^2 + x_2^2)}{(1-x_3)^2} = \frac{4(1-x_3^2)}{(1-x_3)^2} = \frac{4(1+x_3)}{1-x_3}.$$

Solving for x_3 gives

$$x_3 = \frac{y_1^2 + y_2^2 - 4}{y_1^2 + y_2^2 + 4}.$$

Using $y_1 = \frac{2x_1}{1-x_3}$ and $y_2 = \frac{2x_2}{1-x_3}$, we obtain

$$x_1 = \frac{4y_1}{y_1^2 + y_2^2 + 4}, \quad x_2 = \frac{4y_2}{y_1^2 + y_2^2 + 4}.$$

Thus,

$$\phi_1^{-1}(y_1, y_2) = \left(\frac{4y_1}{y_1^2 + y_2^2 + 4}, \frac{4y_2}{y_1^2 + y_2^2 + 4}, \frac{y_1^2 + y_2^2 - 4}{y_1^2 + y_2^2 + 4} \right).$$

The transition function is therefore

$$\phi_2 \circ \phi_1^{-1}(y_1, y_2) = \left(\frac{4y_1}{y_1^2 + y_2^2}, \frac{4y_2}{y_1^2 + y_2^2} \right),$$

which is smooth. Hence, $\{(U_1, \phi_1), (U_2, \phi_2)\}$ forms an atlas for S^2.

(1)(b) The complex projective space $\mathbb{CP}^1$ is the space of complex lines through the origin in $\mathbb{C}^2$. Explicitly, we take $(z_1, z_2) \in \mathbb{C}^2 \setminus \{0\}$ and identify

$$(z_1, z_2) \sim \lambda(z_1, z_2),$$

for any nonzero $\lambda \in \mathbb{C}$. We define

$$U_1 = \{z_1 \neq 0\}, \quad U_2 = \{z_2 \neq 0\},$$

with coordinates

$$\zeta_1 = \frac{z_2}{z_1}, \quad \zeta_2 = \frac{z_1}{z_2}.$$

On the overlap, we have

$$\zeta_1 = \frac{1}{\zeta_2},$$

which is holomorphic. Hence, $\mathbb{CP}^1$ is the Riemann sphere S^2.

(2) Let $X = \{[z_0 : \cdots : z_n] \in \mathbb{CP}^n \mid f(z) = 0\}$, where f is homogeneous of degree d and X is smooth. Work on the standard affine chart $U_0 = \{z_0 \neq 0\} \simeq \mathbb{C}^n$ with coordinates $z = (z_1, \ldots, z_n)$, where the Fubini–Study Kähler potential is

$$K(z, \bar{z}) = \log(1 + |z|^2), \quad |z|^2 = \sum_{k=1}^{n} |z_k|^2,$$

so

$$\omega_{\text{FS}} = \frac{i}{2} \partial\bar{\partial} K, \quad g_{i\bar{j}} = \frac{\partial^2 K}{\partial z_i \partial \bar{z}_j}.$$

Step 1: Reduce to a patch where one coordinate is eliminated. Since X is smooth, at each point of X, at least one partial derivative $\partial f / \partial z_k$ (in homogeneous coordinates) is nonzero. Hence, the open sets $X \cap \{\partial f / \partial z_k \neq 0\}$ cover X. Choose an index (rename it n) and restrict to the open set

$$V := U_0 \cap X \cap \left\{ \frac{\partial f}{\partial z_n} \neq 0 \right\},$$

which is nonempty on the region where we carry out the computation. On V, the implicit function theorem gives a local holomorphic function $z_n = z_n(z_1, \ldots, z_{n-1})$ with $f(z_1, \ldots, z_{n-1}, z_n(z)) \equiv 0$, and hence

$$dz_n = \sum_{i=1}^{n-1} \frac{\partial z_n}{\partial z_i} dz_i, \quad d\bar{z}_n = \sum_{j=1}^{n-1} \frac{\partial \bar{z}_n}{\partial \bar{z}_j} d\bar{z}_j.$$

Step 2: The induced metric determinant on X. Let g^X be the Hermitian metric on X obtained by restricting g to $T^{1,0}X$. A standard linear-algebra computation (Schur complement/Gram determinant) gives, on V,

$$\det(g^X) = \det(g) \, \frac{\sum_{a,b=1}^{n} g^{a\bar{b}} f_a \overline{f_b}}{|f_n|^2}, \tag{6.1}$$

where $f_a := \partial f / \partial z_a$ (affine derivatives on U_0) and $(g^{a\bar{b}}) = (g_{a\bar{b}})^{-1}$. Equivalently, the induced volume form satisfies

$$\frac{(\omega_{\text{FS}}|_X)^{n-1}}{(n-1)!} = \det(g^X) \left(\frac{i}{2}\right)^{n-1} dz_1 \wedge d\bar{z}_1 \wedge \cdots \wedge dz_{n-1} \wedge d\bar{z}_{n-1}.$$

Step 3: Compute the Ricci form of the restricted metric. By definition,

$$\mathscr{R}_{\mathrm{FS}}\big|_X = -\partial\bar{\partial}\log\det(g^X).$$

Using (6.1),

$$\mathscr{R}_{\mathrm{FS}}\big|_X = -\partial\bar{\partial}\log\det(g) \;-\; \partial\bar{\partial}\log\left(\sum_{a,b=1}^{n} g^{a\bar{b}} f_a \overline{f_b}\right) \;+\; \partial\bar{\partial}\log|f_n|^2. \quad (6.2)$$

On U_0, the Fubini–Study determinant is

$$\det(g) = \frac{1}{(1+|z|^2)^{n+1}},$$

so

$$-\partial\bar{\partial}\log\det(g) = (n+1)\,\partial\bar{\partial}\log(1+|z|^2) = \frac{2(n+1)}{i}\,\omega_{\mathrm{FS}}.$$

Step 4: Isolate the $(n+1-d)\,\omega_{\mathrm{FS}}$ term. Define the real function on V:

$$\psi := \log\left(\frac{\sum_{a=0}^{n}\left|\frac{\partial f}{\partial z_a}\right|^2}{(1+|z|^2)^{2(d-1)}}\right).$$

(Here, the derivatives $\partial f/\partial z_a$ are understood in homogeneous coordinates, restricted to U_0 via $z_0 = 1$; this is the same expression used in standard derivations of the adjunction formula.)

Using the explicit inverse Fubini–Study metric

$$g^{a\bar{b}} = (1+|z|^2)(\delta_{ab} + z_a\bar{z}_b) \quad (a,b = 1,\ldots,n),$$

together with the Euler homogeneity on X, namely $\sum_{a=0}^{n} z_a \frac{\partial f}{\partial z_a} = d\,f = 0$ on X, one obtains on V the identity

$$\sum_{a,b=1}^{n} g^{a\bar{b}} f_a \overline{f_b} = (1+|z|^2)^{d-1}\,e^{\psi}.$$

Substituting this into (6.2) gives

$$\begin{aligned}
\mathscr{R}_{\mathrm{FS}}\big|_X &= \frac{2(n+1)}{i}\,\omega_{\mathrm{FS}} - \partial\bar{\partial}\log\big((1+|z|^2)^{d-1}e^{\psi}\big) + \partial\bar{\partial}\log|f_n|^2 \\
&= \frac{2(n+1)}{i}\,\omega_{\mathrm{FS}} - (d-1)\,\partial\bar{\partial}\log(1+|z|^2) - \partial\bar{\partial}\psi + \partial\bar{\partial}\log|f_n|^2 \\
&= \frac{2(n-d+1)}{i}\,\omega_{\mathrm{FS}} - \partial\bar{\partial}(\psi - \log|f_n|^2).
\end{aligned}$$

The last term is of the form $\partial\bar{\partial}(\cdots)$, i.e., an exact $(1,1)$-form ("total derivatives"). Thus,

$$\boxed{\mathscr{R}_{\text{FS}}\big|_X = \frac{2(n-d+1)}{i}\,\omega_{\text{FS}} + (\text{total derivatives}).}$$

Step 5: Existence of a Ricci-flat metric in the same class. Taking cohomology classes annihilates $\partial\bar{\partial}$-exact terms, and hence

$$[\mathscr{R}_{\text{FS}}|_X] = \frac{2(n-d+1)}{i}\,[\omega_{\text{FS}}|_X].$$

In particular, if $d = n+1$, then $[\mathscr{R}_{\text{FS}}|_X] = 0$, i.e., $c_1(X) = 0$.

By Yau's theorem, every Kähler class on a compact Kähler manifold with $c_1 = 0$ contains a unique Ricci-flat Kähler metric. Therefore, when $d = n+1$, there exists a Ricci-flat Kähler metric in the same Kähler class as $\omega_{\text{FS}}|_X$.

6.2 Tutorial 2: Numerical Differential Equations on Complex Manifolds

As we move from analytic metrics on complex manifolds to numerical metrics on Calabi–Yau manifolds, it is useful to first introduce the computational methods that will be employed. As a warm-up towards Calabi–Yau metrics, this tutorial considers a simplified version of the full problem: solving a system of differential equations on complex projective space $\mathbb{CP}^1$ using stochastic gradient descent (SGD).

The concepts introduced here will be developed further in the following tutorial, where each component of the problem is replaced appropriately to address this chapter's primary goal. Schematically,

$$\begin{array}{ccc}
\{\text{differential equations}\} & \text{on } \{\mathbb{CP}^1\} \text{ with} & \{\text{SGD}\} \\
\downarrow & \downarrow & \downarrow \\
\{R = 0\} & \text{on } \{CY_3\} \text{ with} & \{\text{ML}\}.
\end{array}$$

The problem in this tutorial is to find a global function F on the manifold $\mathbb{CP}^1$, presented in homogeneous coordinates (z_0, z_1), which satisfies the conditions

$$\frac{\partial F}{\partial z_0} = \frac{\bar{z}_1(|z_0|^2 + |z_1|^2 - \bar{z}_0)}{(|z_0|^2 + |z_1|^2)^2}, \tag{6.3}$$

$$\frac{\partial F}{\partial \bar{z}_0} = \frac{-z_0^2 \bar{z}_1}{(|z_0|^2 + |z_1|^2)^2}, \tag{6.4}$$

$$\frac{\partial F}{\partial z_1} = \frac{-z_0 \bar{z}_1^2}{(|z_0|^2 + |z_1|^2)^2}, \tag{6.5}$$

$$\frac{\partial F}{\partial \bar{z}_1} = \frac{z_0(|z_0|^2 + |z_1|^2 - z_1)}{(|z_0|^2 + |z_1|^2)^2}, \tag{6.6}$$

where $\bar{z}$ denotes the complex conjugate of z. To solve this problem, we assume an *ansatz* solution f of the form

$$f(z_0, \bar{z}_0, z_1, \bar{z}_1) = \frac{c_{1,11} z_0 \bar{z}_0 + c_{1,10} z_0 \bar{z}_1 + c_{1,01} z_1 \bar{z}_0 + c_{1,00} z_1 \bar{z}_1}{|z_0|^2 + |z_1|^2}, \tag{6.7}$$

where the constant real coefficients $c_{\alpha,\beta\gamma}$ are indexed such that for each monomial term $c_{\alpha,\beta\gamma} z_i \bar{z}_j$, the indices satisfy

$$\alpha = \beta + i = \gamma + j.$$

This indexing convention allows for straightforward generalization to higher-order *ansätze* of the form $c_{\alpha,\beta\gamma} z_i^n \bar{z}_j^m$, satisfying

$$\alpha = \beta + in = \gamma + jm.$$

The coefficients are then optimized using SGD in the $\mathbb{R}^4$ coefficient space, minimizing a loss function constructed from the differentiability conditions. To implement this procedure, a `sagemath` Jupyter notebook is provided at

`https://github.com/andrei-const/ML_MathsPhys_Oxford2023/blob/main/Tutorial2.ipynb.`

For the symbolic manipulation, a `sagemath` kernel is required (a standard `python` kernel is insufficient). In general, `sagemath` is a `python` extension providing a wide range of mathematical functionality. It can be installed locally from https://sagemath.org or used remotely via the `cocalc` platform at https://cocalc.com. We now follow the structure of the notebook, explaining the role of each cell.

[1] The first cell imports the required libraries and defines a random seed. This pseudo-random number allows stochastic runs to be reproduced identically. The seed may be set to `None` or changed arbitrarily to explore the stochastic nature of the optimization.

```
# Import libraries
import numpy as np
import matplotlib.pyplot as plt

# Set the random seed (None for no seeding)
seed = 3
```

[2] The second cell samples points on $\mathbb{CP}^1$ and defines the train:validation: test splits. The number of sampled points is controlled by `number_of_points`. Increasing this improves accuracy but slows computation. Random four-vectors are sampled from a normal distribution, normalized to lie on the real three-sphere S^3, and interpreted as $(\text{Re } z_0, \text{Im } z_0, \text{Re } z_1, \text{Im } z_1)$. Example points may be viewed with `homogeneous_points[:10]`. The dataset is then divided according to `split_proportions`.

```
# Point sampling on P1
number_of_points = int(1e3)

# Homogeneous points
if seed:
    np.random.seed(seed)

sphere_points = np.random.normal(loc=0.0, scale=1.0,
                                 size=(number_of_points, 4))

sphere_points /= np.linalg.norm(sphere_points, axis=1)[:, np.newaxis]

homogeneous_points = np.array([
    [pt_set[0] + 1j*pt_set[1],
     pt_set[2] + 1j*pt_set[3]]
    for pt_set in sphere_points
])

del sphere_points

# Split into train, validation, test
split_proportions = [0.7, 0.1, 0.2]
# note: only uses train and validation proportions,
# assumes test as the remainder

train_points = homogeneous_points[
    :int(np.floor(number_of_points * split_proportions[0]))
]
```

```
validation_points = homogeneous_points[
    int(np.floor(number_of_points * split_proportions[0])):
    int(np.floor(number_of_points *
        (split_proportions[0] + split_proportions[1])))
]

test_points = homogeneous_points[
    int(np.floor(number_of_points *
        (split_proportions[0] + split_proportions[1]))):
]
```

[3] The third cell defines the *ansatz* function f as in (6.7). The maximum order is initially set to one via `max_order` but may be increased to include higher-power monomials. Independent conjugate variables are introduced so that partial differentiation is explicit. The *ansatz* is constructed iteratively, with a general matrix of real coefficients multiplying the homogeneous monomial terms. The final expression is printed as output.

```
# Define the ansatz for the function in homogeneous coordinates
max_order = 1   # select the order to extend the ansatz to

z_0, zb_0, z_1, zb_1 = var('z_0, zb_0, z_1, zb_1', domain='complex')
    # define the homogeneous variables
    # (note: define conjugates as separate variables
    # so that partial differentiation is possible)

# Initialise the function
f(z_0, zb_0, z_1, zb_1) = 0

# Loop through the degree orders
# (following cells currently only work for max_order = 1)
for i in range(1, max_order + 1):

    # Define the coefficients for this order as variables
    c = matrix(
        SR, i+1, i+1,
        lambda alpha, beta:
            var('c_{}_{}_{}'.format(i, alpha, beta), domain='real')
    )
    # print('Coefficients:', c)

    # Initialise the function contribution for this order
    fi(z_0, z_1) = 0

    # Loop through all homogeneous polynomials in both
    # holomorphic and antiholomorphic parts,
```

```
    # and add them with their respective general coefficient
    for j1 in range(i+1):
        for j2 in range(i+1):
            fi += (
                c[j1, j2]
                * z_0**j1 * z_1**(i-j1)
                * zb_0**j2 * zb_1**(i-j2)
                / (z_0*zb_0 + z_1*zb_1)**i
            )

    # Add this order's ansatz to the full ansatz
    f += fi

# Print the final ansatz form
print(f.full_simplify())
```

[4] The fourth cell implements the differentiability conditions (6.3)–(6.6) and defines the loss function. Each condition contributes a term measuring the normed difference between the condition and the corresponding derivative of the ansatz. These components are summed to define the total loss. A function `Loss` evaluates this quantity for trial coefficients and sample points. A complementary function computes the gradient of the loss with respect to each coefficient. Example evaluations are provided for testing.

```
# Define the differentiability conditions
dFdz0(z_0, z_1) = (z_1*conjugate(z_1)**2) /
        (z_0*conjugate(z_0) + z_1*conjugate(z_1))**2

dFdzb0(z_0, z_1) = (-z_0**2*conjugate(z_1)) /
        (z_0*conjugate(z_0) + z_1*conjugate(z_1))**2

dFdz1(z_0, z_1) = (-z_0*conjugate(z_1)**2) /
        (z_0*conjugate(z_0) + z_1*conjugate(z_1))**2

dFdzb1(z_0, z_1) = (z_0**2*conjugate(z_0)) /
        (z_0*conjugate(z_0) + z_1*conjugate(z_1))**2

# Compute symbolic loss function
LossA = norm(f.diff(z_0).substitute(zb_0=conjugate(z_0),
                            zb_1=conjugate(z_1)) - dFdz0)

LossB = norm(f.diff(zb_0).substitute(zb_0=conjugate(z_0),
                            zb_1=conjugate(z_1)) - dFdzb0)
```

```
LossC = norm(f.diff(z_1).substitute(zb_0=conjugate(z_0),
                          zb_1=conjugate(z_1)) - dFdz1)

LossD = norm(f.diff(zb_1).substitute(zb_0=conjugate(z_0),
                           zb_1=conjugate(z_1)) - dFdzb1)

Total_Loss = LossA + LossB + LossC + LossD
Total_Loss = Total_Loss.full_simplify()
# print(Total_Loss)

# Define the loss function
def Loss(constants, point):
    return float(
        real_part(
            Total_Loss.substitute(
                c_1_0_0=constants[0],
                c_1_0_1=constants[1],
                c_1_1_0=constants[2],
                c_1_1_1=constants[3]
            ).substitute(
                z_0=point[0],
                z_1=point[1]
            )
        )
    )

# Compute symbolic loss gradient wrt coefficients
gradient_fn = [
    Total_Loss.diff(c_1_0_0).full_simplify(),
    Total_Loss.diff(c_1_0_1).full_simplify(),
    Total_Loss.diff(c_1_1_0).full_simplify(),
    Total_Loss.diff(c_1_1_1).full_simplify()
]

# Define the loss gradient function
# (evaluating for the given coefficients at the considered point)
def LossGradient(constants, point):

    gradient_vec = [
        coeff_grad.substitute(
            c_1_0_0=constants[0],
            c_1_0_1=constants[1],
            c_1_1_0=constants[2],
            c_1_1_1=constants[3]
        )
```

```
        for coeff_grad in gradient_fn
    ]

    gradient = np.real(
        np.array([
            coeff_grad.substitute(
                z_0=point[0],
                z_1=point[1]
            )
            for coeff_grad in gradient_vec
        ], dtype=complex)
    )
    return gradient

# Test
example_coeffs, example_point = [-1, 2, 3, 4], [3*I, 5+I]

print(f'Example coeffs {example_coeffs}, and point {example_point}')
print(f'Loss: {Loss(example_coeffs, example_point)}')
print(f'Loss gradient: {LossGradient(example_coeffs,
  example_point)}')
```

[5] The fifth cell runs the SGD algorithm. The coefficient vector is initialized, and the hyperparameters are set: `batch_size`, `number_of_epochs`, `learning_rate`, `momentum`, and `tolerance`. These parameters control the optimization behaviour and may be tuned. At each epoch, the datasets are shuffled, batches are processed, losses and gradients are computed and averaged, and the coefficients are updated. Optional formulas for learning-rate decay and momentum are provided. Training and validation losses are recorded after each epoch. Upon completion, the cell outputs the final loss and coefficient values. If the loss is sufficiently small, the coefficients provide an accurate approximation to the true solution F.

```
# Initialise coefficients
if seed:
    np.random.seed(seed)

coefficients = np.random.normal(loc=0.0, scale=3.0, size=4)
print(f'Coefficients: {coefficients}\n')

# Define SGD hyperparameters
batch_size = 32
number_of_epochs = 10
learning_rate = 1e-2
momentum = 0.8
tolerance = 1e-8
```

```
# Perform the SGD
epoch_losses = []

for epoch in range(number_of_epochs):

    # Shuffle the data (so batches change each epoch)
    np.random.shuffle(train_points)
    np.random.shuffle(validation_points)

    # Loop over batches
    print(f'Epoch: {epoch+1}\nProgress: ', end='')
    batches_losses = []
    coeff_step = 0

    for batch_count, batch_idx in enumerate(
        range(int(np.floor(len(train_points)/batch_size)))
    ):

        # Compute loss for the batch
        batches_losses.append(
            np.mean([
                Loss(coefficients, train_points[idx])
                for idx in range(
                    batch_idx*batch_size,
                    min(number_of_points+1,
                        (batch_idx+1)*batch_size)
                )
            ])
        )

        # Compute the gradient across the batch
        grad = np.mean([
            LossGradient(coefficients, train_points[idx])
            for idx in range(
                batch_idx*batch_size,
                min(number_of_points+1,
                    (batch_idx+1)*batch_size)
            )
        ], axis=0)

        # Gradient descent step
        coeff_step = -learning_rate * grad

        # Optional variants:
        # coeff_step = ((number_of_epochs-epoch)/number_of_epochs)
```

```
        #               * learning_rate * grad       # with decay
        # coeff_step = momentum*coeff_step - learning_rate*grad
        # coeff_step = momentum*coeff_step -
        #               ((number_of_epochs-epoch)/number_of_epochs)
        #               * learning_rate * grad

        # Early stopping condition
        if np.all(np.abs(coeff_step) <= tolerance):
            print('...early stopping', end=' ')
            break

        # Update the coefficients
        coefficients += coeff_step

        if batch_count % int(len(train_points)/(20*batch_size)) == 0:
            print('=', end='')

    print('x')

    # Compute epoch losses
    train_loss = np.mean(batches_losses)
    validation_loss = np.mean([
        Loss(coefficients, val_point)
        for val_point in validation_points
    ])

    print(
        f'Training Loss: {train_loss}\n'
        f'Validation Loss: {validation_loss}\n'
        f'Coefficients: {coefficients}\n'
    )

    epoch_losses.append([train_loss, validation_loss])


# Run testing
epoch_losses = np.array(epoch_losses)

test_loss = np.mean([
    Loss(coefficients, test_point)
    for test_point in test_points
])
```

```
print(
    f'##########\n'
    f'Final Test Loss: {test_loss}\n'
    f'Final Coefficients: {np.round(coefficients)}'
)
```

[6] The sixth cell plots the training and validation losses over the epochs. If only one epoch is used, batch losses are plotted instead. Successful convergence is indicated by both losses approaching zero.

```
# Loss plotting
plt.figure()

if number_of_epochs == 1:
    plt.plot(
        range(1, len(batches_losses) + 1),
        batches_losses,
        label='Batch')
else:
    plt.plot(
        range(1, len(epoch_losses) + 1),
        epoch_losses[:, 0],
        label='Train')
    plt.plot(
        range(1, len(epoch_losses) + 1),
        epoch_losses[:, 1],
        label='Validation')

plt.title('Losses')
plt.xlabel('Epoch')
plt.ylabel('Loss')
plt.ylim(0)
plt.xlim(0)
plt.legend(loc='best')
plt.grid()
plt.tight_layout()
# plt.savefig('Losses.pdf')
```

Running the notebook sequentially demonstrates a successful application of SGD to this problem. The final values of the coefficients $(c_{1,11}, c_{1,10}, c_{1,01}, c_{1,00})$ determine the function F.

Once familiar with this procedure, numerous extensions may be explored, as listed in the notebook and summarized as follows:

- Investigate the effect of different sampling strategies and sample sizes.
- Extend SGD with alternative learning rates, decay schemes, momentum, RMSProp, and Adam.
- Modify the loss function to higher-order norms, introduce regularization, or explore alternative loss structures.
- Construct losses based on second-order derivatives, closer to the Ricci-scalar condition.
- Solve for alternative target functions on $\mathbb{CP}^1$ using the same *ansatz*.

With experimentation, one observes that the solution is not unique. This reflects a symmetry in the differentiability conditions: adding constants to the *ansatz* does not change the derivatives. A natural canonical choice can be selected by introducing suitable regularization into the loss function.

6.3 Tutorial 3: Ricci-Flat Metrics on Calabi–Yau Threefolds: ML Approach

6.3.1 *Part 1: A toy model*

In the first section, we write a toy model from scratch to better understand the problem. Once you have the `cymetric` package installed, the necessary libraries can be imported using the following:

```
import numpy as np
import tensorflow as tf
import matplotlib.pyplot as plt
from cymetric.pointgen.pointgen import PointGenerator
```

Here, we also define the function `convert_to_hermitian`, which converts the outputs of the neural network (real vectors in $\mathbb{R}^9$) to 3×3 complex hermitian matrices. This will be useful later.

```
def convert_to_hermitian(x,nfold=3):
    t1=f.reshape(tf.complex(x,tf.zeros_like(x)),(-1,nfold,nfold))
    up=tf.linalg.band_part(t1,0,-1)
    low=tf.linalg.band_part(1j*t1,-1,0)
    out=up+tf.transpose(up,perm=[0,2,1])-tf.linalg.band_part(t1,0,0)
    return out+low+tf.transpose(low,perm=[0,2,1],conjugate=True)
```

For this example, we work out the simplest case: the Fermat quintic

$$z_1^5 + z_2^5 + z_3^5 + z_4^5 + z_5^5 = 0.$$

The input data X used for training will be the points on the quintic. They can be sampled using the point generator provided by the `cymetric` package, which can be defined with the following format:

```
monomials = 5*np.eye(5, dtype=np.int64)
coefficients = np.ones(5)
kmoduli = np.ones(1)
ambient = np.array([4])
pg = PointGenerator(monomials, coefficients, kmoduli, ambient)
```

The points can now be sampled:

```
points = pg.generate_points(1000)
# Keep points which have z_5 = 1
points = points[np.abs(points[:,-1] - 1) < 0.001]
```

Next, we need to compute the holomorphic volume form $\Omega \wedge \bar{\Omega}$. The holomorphic n-form Ω is defined as $1/(\partial f/\partial z_i)$, where z_i can be chosen as any affine coordinate. Later on, this coordinate will be eliminated when restricting the metric to the Calabi–Yau manifold. In practice, however, the denominator being too small may lead to numerical instability. To avoid this, z_i is usually chosen so that $\partial f/\partial z_i$ is the largest at each point. This is done automatically in the modern Calabi–Yau metric packages, but it will be a little bit messy to implement from scratch for our toy example. Therefore, for pedagogical reasons, we make another simplification here: we manually set z_i to be z_4. Since only the component of the volume form is needed in the training process, once the n-form Ω is computed, we can simply take the conjugation and multiply them together:

```
Omega = 1/(5*points[:,0]**4)
Omegabar = np.conj(Omega)
holomorphic_volume_form = Omega * Omegabar
holomorphic_volume_form[0:10]
```

Before we feed our data to the neural network, we also need to convert the data points to real values. In principle, a complex network can also do the job (see the holomorphic neural network in the `MLGeometry` package), but real numbers are just easier to deal with:

```
points_real=tf.concat((tf.math.real(points),tf.math.imag(points)),
                                                        axis=1)
```

Now that we have all of our data ready, the next step is to construct a neural network to approximate the metric. Here, we pick a simple three-layer architecture without any hyperparameter tuning. The neural network works essentially as a function, mapping the points on the manifolds to the elements of 3×3 complex matrices. But since the metric is Hermitian, there only needs to be nine neurons in the output layer:

```
NNmetric = tf.keras.models.Sequential()
NNmetric.add(tf.keras.layers.Dense(64,activation='relu',
                                   input_shape=(10,)))
NNmetric.add(tf.keras.layers.Dense(32, activation='relu'))
NNmetric.add(tf.keras.layers.Dense(9))
```

As discussed in the previous chapter, the Ricci-flat condition on the Calabi–Yau manifold is equivalent to the equality of two volume forms. One of these forms can be computed by taking the determinant of the metric, and the other one is the volume form we have just computed. Thus, the loss function can be defined as

$$\int \mathrm{d}\mu \left|1 - \frac{\det g}{\kappa \Omega \wedge \bar{\Omega}}\right| \approx \frac{V}{N} \sum \left|1 - \frac{\det g}{\kappa \Omega \wedge \bar{\Omega}}\right| .$$

In order for the right-hand side to approximate well the integral, the sum needs to be weighted by the correct measure, as discussed around Eq. (5.23). However, for the purpose of simplicity, we ignore this aspect here and also set the constant κ to 1:

```
def loss_func(NN_output, holomorphic_volume_form):
    g = convert_to_hermitian(NN_output)
    omega = tf.linalg.det(g)
    loss = tf.reduce_mean(tf.abs(1-omega/holomorphic_volume_form))
    return loss
```

Now, with the data, network, and the loss function ready, we can train it just like we do in usual ML problems and plot the training history in Figure 6.1:

```
epoch = 0
max_epoch = 300
optimizer = tf.keras.optimizers.Adam()
loss_history = []

while epoch < max_epoch:
    with tf.GradientTape() as tape:
        loss=loss_func(NNmetric(points_real),holomorphic_volume_form)
    grads = tape.gradient(loss, NNmetric.trainable_weights)
    optimizer.apply_gradients(zip(grads, NNmetric.trainable_weights))
    loss_history.append(loss)
    epoch = epoch + 1
plt.plot(loss_history)
```

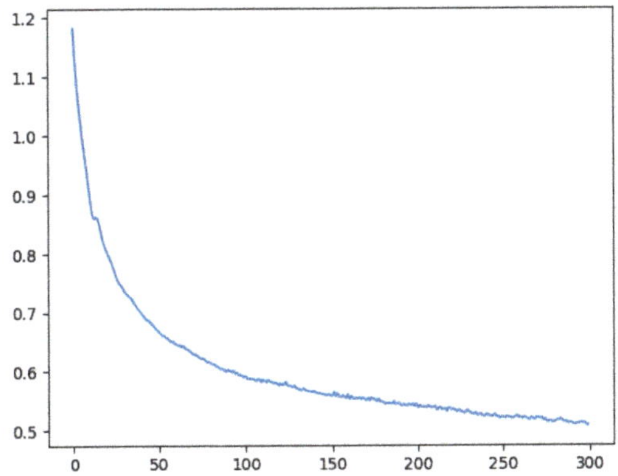

Fig. 6.1. Loss function during training. Horizontal axis: number of epochs.

As the plot indicates, the loss goes down quickly after only a few epochs of training. Hopefully, this toy example gives a rough idea about how numerical Calabi–Yau metrics can be obtained with ML techniques. Note, however, that multiple simplifications have been made, which would need to be corrected, as indeed, has been done in modern computer packages. In the following two sections, we demonstrate how to use two of these packages: `cymetric` and `MLGeometry`.

6.3.2 *Part 2: The Cymetric package*

In this section, we compute the full-fledged Calabi–Yau metrics with the `cymetric` package. We begin with the following:

```
import os as os
import pickle as pickle
import matplotlib.pyplot as plt
import seaborn as sns
sns.set_style("darkgrid")

import pandas as pd
import numpy as np
import tensorflow as tf

from cymetric.pointgen.pointgen import PointGenerator
from cymetric.models.tfhelper import prepare_tf_basis, train_model
from cymetric.models.callbacks import RicciCallback, SigmaCallback,
        VolkCallback, KaehlerCallback, TransitionCallback
from cymetric.models.tfmodels import MultFSModel, PhiFSModel
from cymetric.models.metrics import SigmaLoss, KaehlerLoss,
        TransitionLoss, VolkLoss, RicciLoss, TotalLoss
```

The point generator is defined as usual:

```
monomials = 5*np.eye(5, dtype=np.int64)
coefficients = np.ones(5)
kmoduli = np.ones(1)
ambient = np.array([4])
pg = PointGenerator(monomials, coefficients, kmoduli, ambient)
```

Next, we generate points using the `prepare_dataset` method, which also deals with the patchwork and computes the holomorphic volume form automatically. It returns the constant κ. The `prepare_basis` method generates useful information; for instance, it automatically finds the coordinate that has the largest $\partial f/\partial z_i$:

```
dirname = 'fermat_pg'
n_p = 10000

kappa = pg.prepare_dataset(n_p, dirname)
pg.prepare_basis(dirname, kappa=kappa)
data = np.load(os.path.join(dirname, 'dataset.npz'))

BASIS=np.load(os.path.join(dirname,'basis.pickle'),allow_pickle=True)
BASIS=prepare_tf_basis(BASIS)
```

We also define a number of callbacks that measure the performance of our model during training. `KaehlerCallback`, for example, is used to measure how close the metric is to a Kähler metric; `TransitionCallback` is used to measure the difference between the metrics obtained on different patches, etc.

```
ricci_cb = RicciCallback((data['X_val'], data['y_val']),
            data['val_pullbacks'])
sigma_cb = SigmaCallback((data['X_val'], data['y_val']))
volk_cb = VolkCallback((data['X_val'], data['y_val']))
kaehler_cb = KaehlerCallback((data['X_val'], data['y_val']))
transition_cb = TransitionCallback((data['X_val'], data['y_val']))
cb_list = [ricci_cb, sigma_cb, volk_cb, kaehler_cb, transition_cb]
```

Once again, we can set up a neural network and use it to construct the metric. In the previous toy example, we used a "free" model, i.e., the network directly produced the components of the metric, which is straightforward but not efficient. Since it is assumed that the Ricci-flat metric on the Calabi–Yau manifold is close to the Fubini–Study metric, we can let the network represent a perturbation on top of the Fubini–Study metric. Here, we construct the metric with the built-in class `MultFSModel`, which uses the *ansatz* $g = g_{FS}(1 + g_{NN})$:

```
nlayer = 3
nHidden = 64
act = 'gelu'
nEpochs = 50
# The batch size, the second one is for VolkLoss Specifically
bSizes = [64, 50000]
# weights on the loss functions
alpha = [1., 1., 1., 1., 1.]
nfold = 3
n_in = 2*5
n_out = nfold**2

nn = tf.keras.Sequential()
nn.add(tf.keras.Input(shape=(n_in)))
for i in range(nlayer):
    nn.add(tf.keras.layers.Dense(nHidden, activation=act))
nn.add(tf.keras.layers.Dense(n_out, use_bias=False))

fmodel = MultFSModel(nn, BASIS, alpha=alpha)
```

We have imported several predefined loss functions at the beginning, which can now be used:

```
cmetrics = [TotalLoss(), SigmaLoss(), KaehlerLoss(),
            TransitionLoss(), VolkLoss(), RicciLoss()]
```

Finally, we can train the model with the `train_model` function:

```
opt = tf.keras.optimizers.Adam()
fmodel, training_history = train_model(fmodel, data, optimizer=opt,
        epochs=nEpochs, batch_sizes=[64, 50000],
        verbose=1, custom_metrics=cmetrics, callbacks=cb_list)
```

We then plot the results, as shown in Figure 6.2:

```
hist = training_history
hist['epochs'] = list(range(nEpochs))
hist_data = pd.DataFrame(hist)
fig, [[ax1, ax2], [ax3, ax4]] = plt.subplots(2, 2, figsize=(10,10))
ax1.set_title("Monge-Ampere loss")
sns.scatterplot(x='epochs', y='sigma_loss', data=hist_data,
        alpha=0.75, ax=ax1, label="train")
sns.scatterplot(x='epochs', y='sigma_val', data=hist_data,
        alpha=0.75, ax=ax1, label="val")
```

```
ax2.set_title("Kahler loss")
sns.scatterplot(x='epochs', y='kaehler_loss', data=hist_data,
       alpha=0.75, ax=ax2, label="train")
sns.scatterplot(x='epochs', y='kaehler_val', data=hist_data,
       alpha=0.75, ax=ax2, label="val")

ax3.set_title("Transition loss")
sns.scatterplot(x='epochs', y='transition_loss', data=hist_data,
       alpha=0.75, ax=ax3, label="train")
sns.scatterplot(x='epochs', y='transition_val', data=hist_data,
       alpha=0.75, ax=ax3, label="val")

ax4.set_title("Ricci measure")
sns.scatterplot(x='sigma_val', y='ricci_val', data=hist_data,
       alpha=0.75, ax=ax4, label="test");
```

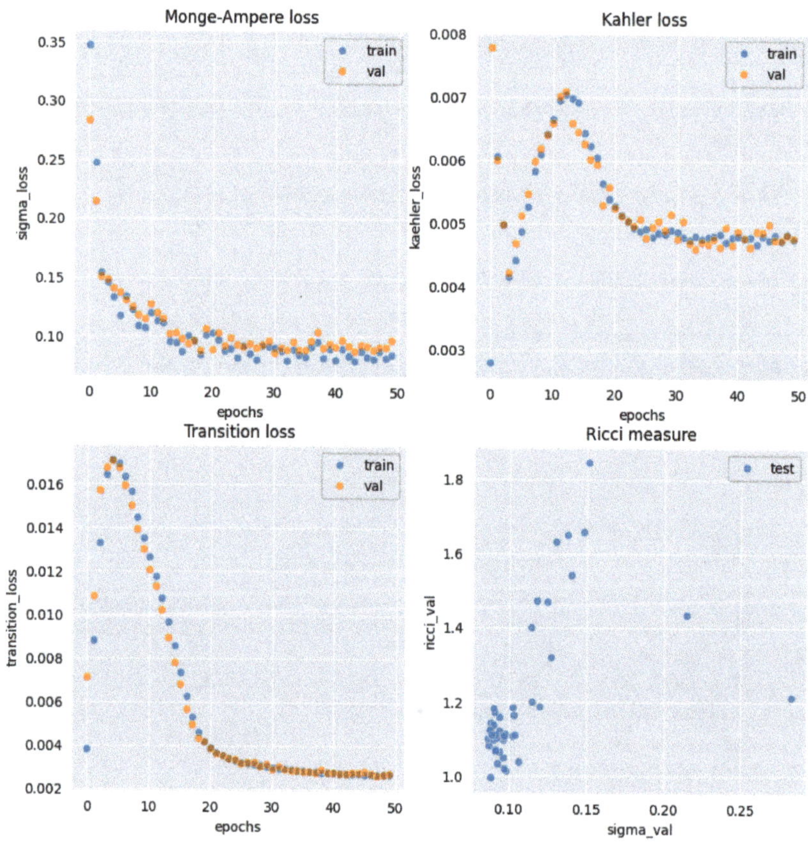

Fig. 6.2. Loss functions during training.

The neural network has not been fine-tuned yet, but the outcomes already surpass those of the toy model used earlier. The metric is also better defined. The model can be further improved by tuning the hyperparameters (depth and width of the network, learning rate, etc.) One can also experiment with different hypersurfaces, architectures, and loss functions.

6.3.3 *Part 3: The MLGeometry package*

In this section, we introduce the package `MLGeometry`, which uses a different approach to resolving the challenges encountered in the toy model.

The notebook version of this tutorial is hosted on Github:

`https://github.com/yidiq7/MLGeometry/blob/master/Guide.ipynb`

We start by importing the necessary packages:

```
import sympy as sp
import tensorflow as tf
import tensorflow.python.keras.backend as K
import tensorflow_probability as tfp

import MLGeometry as mlg
from MLGeometry import bihomoNN as bnn

import math
import numpy as np
import matplotlib.pyplot as plt
```

Optionally, you can set your own random seeds in order to get reproducible results:

```
np.random.seed(42)
tf.random.set_seed(42)
```

The way to define a hypersurface and generate points is slightly different in `MLGeometry`. We first define the coordinates symbolically using `sympy` and write down the expression of the hypersurface explicitly:

```
z0, z1, z2, z3, z4 = sp.symbols('z0, z1, z2, z3, z4')
Z = [z0,z1,z2,z3,z4]
f = z0**5 + z1**5 + z2**5 + z3**5 + z4**5 + 0.5*z0*z1*z2*z3*z4
```

We then use the `Hypersurface` class in the `mlg.hypersurface` module. The parameter `n_pairs` is the number of random pairs of points used to form the random lines in $\mathbb{CP}^{N+1}$. By Bezout's theorem, each line intersects the hypersurface in precisely d points, where d is the number of homogeneous coordinates. Therefore, the total number of points is `d * n_pairs`. We define both the training set and testing set here:

```
n_pairs = 10000
HS_train = mlg.hypersurface.Hypersurface(Z, f, n_pairs)
HS_test = mlg.hypersurface.Hypersurface(Z, f, n_pairs)
```

Similar to the `prepare_dataset` function in `cymetric`, there is a function called `mlg.tf_dataset.generate_dataset` in `MLGeometry`. This generates the corresponding Tensorflow dataset for the hypersurface, which has four components: the points on the hypersurface, the holomorphic volume form $\Omega \wedge \bar{\Omega}$, the mass reweighting the points distribution, and the restriction of the Kähler metric to a subpatch. The restriction contains an extra linear transformation so that points on different affine patches can all be processed in one call. It is also possible to generate a dataset only on one specific affine patch:

```
train_set = mlg.tf_dataset.generate_dataset(HS_train)
test_set = mlg.tf_dataset.generate_dataset(HS_test)

# Shuffle and batch
train_set = train_set.shuffle(HS_train.n_points).batch(1000)
test_set = test_set.shuffle(HS_test.n_points).batch(1000)
```

```
class Kahler_potential(tf.keras.Model):
    def __init__(self):
        super(Kahler_potential, self).__init__()
        # The first layer transforms the complex points to the
                bihomogeneous form.
        # The number of the outputs is d^2, where d is the number of
                coordinates.
        self.bihomogeneous = bnn.Bihomogeneous(d=5)
        self.layer1 = bnn.SquareDense(5**2, 70, activation=tf.square)
        self.layer2 = bnn.SquareDense(70, 100, activation=tf.square)
        self.layer3 = bnn.SquareDense(100, 1)

    def call(self, inputs):
        x = self.bihomogeneous(inputs)
        x = self.layer1(x)
        x = self.layer2(x)
        x = self.layer3(x)
        x = tf.math.log(x)
        return x

model = Kahler_potential()
```

```
@tf.function
def volume_form(points, Omega_Omegabar, mass, restriction):

    kahler_metric = mlg.complex_math.complex_hessian
                (tf.math.real(model(points)), points)
    volume_form = tf.matmul(restriction,
               tf.matmul(kahler_metric, restriction, adjoint_b=True))
    volume_form = tf.math.real(tf.linalg.det(volume_form))

    # Calculate the normalization constant to make the overall
                integration as 1
    # It is a batchwise calculation but we expect it to converge
                to a constant eventually
    weights = mass / tf.reduce_sum(mass)
    factor = tf.reduce_sum(weights * volume_form / Omega_Omegabar)

    return volume_form / factor
```

```
optimizer = tf.keras.optimizers.Adam()
loss_func = mlg.loss.weighted_MAPE
```

```
max_epochs = 500
epoch = 0
while epoch < max_epochs:
    epoch = epoch + 1
    for step, (points, Omega_Omegabar, mass, restriction)
                    in enumerate(train_set):
        with tf.GradientTape() as tape:
            det_omega = volume_form(points, Omega_Omegabar,
                                mass, restriction)
            loss = loss_func(Omega_Omegabar, det_omega, mass)
            grads = tape.gradient(loss, model.trainable_weights)
        optimizer.apply_gradients(zip(grads, model.trainable_weights))
    if epoch % 50 == 0:
        print("epoch %d: loss = %.5f" % (epoch, loss))
```

PART 4

REINFORCEMENT LEARNING FOR KNOT THEORY

The first lecture introduces reinforcement learning (RL) and how it can be used to obtain verifiable results in mathematics. In the second lecture, we explain how to apply this to problems in knot theory, involving the questions of whether a knot is trivial or not and whether it has a property called sliceness, which is important for the smooth Poincare conjecture in four dimensions. In the third lecture, we developed an RL agent based on the IMPALA architecture that often finds minimal unknotting trajectories for a knot diagram up to 200 crossings. We use this to determine the unknotting number of 57,000 knots and other applications.

CHAPTER 7

INTRODUCTION TO COMPUTATIONAL COMPLEXITY AND REINFORCEMENT LEARNING

FABIAN RUEHLE

7.1 Introduction

As we have seen in Section 1.1, we can use first-order methods such as gradient descent to find extrema in a very complicated, billion-dimensional parameter space. With the advent of the modern machine learning (ML) era, rapid, parallelized, easy-to-use code methods have been developed to perform this task. This makes it tempting to try the following strategy: formalize the problem you want to solve as an extremization task and solve it with gradient descent. While this can be a good strategy, it does not always work. Most notably, if the problem is discrete, one cannot use continuous optimization methods such as gradient descent, which requires derivatives.

Discrete optimization problems are ubiquitous. They appear in discrete mathematics in group theory, number theory, graph theory, topology, etc. In physics, they appear in any quantum theory since quantization means making a continuum discrete.[1] Another prominent example for a discrete optimization task is the board game Go. Go is a two-player game in which players take turns and place stones on a 19×19 grid with the

[1]This list goes on for other STEM fields; protein folding in biology is another very prominent example.

aim of surrounding more area with their stones than their opponent. In 2016, Deepmind (now owned by Google) developed an algorithm called Alpha Go [1], based on an ML technique called reinforcement learning (RL), which was beating the world champion in Go (Lee Sedol). Since then, Google has improved their algorithm with the publication of AlphaGo Zero [2], trained with zero human knowledge (which is why it is called AlphaGo *Zero*), the only input being the rules of the game. This algorithm managed to beat the original AlphaGo 100-0, demonstrating that RL can achieve superhuman performance on discrete optimization tasks.

7.2 Lecture 1

7.2.1 *Introduction to computational complexity*

All the problems mentioned above are of the same type, which one might call combinatorially complex: one is given some object like a Go board, a quantized potential, a Diophantine equation (which is a multivariate polynomial equation with integer coefficients and variables that can only take integer values), a protein, etc., and the task is to optimize it in some way. For example, find a sequence of moves that wins the Go game, or find the number of quanta that minimizes the potential. On this object, there are multiple discrete manipulations that can be performed to solve the task. This number need not be big; in the example of Go, each player can place one stone at an empty grid point at their turn, meaning they have at most $19 \times 19 = 361$ possible actions they can perform. If one finds the "correct" sequence of manipulations that solves the task, one can answer the question whether the given problem has a solution in the affirmative. The problem is that, even if the number of actions is small, the number of possible states these actions can lead to is often very large. In the example of Go, choosing a sequence of 361 actions of where to put stones on the board, one can obtain $3^{361} \approx 10^{170}$ different[2] Go boards. Compared to such (exponentially) big numbers, the estimated 10^{80} atoms in the observable universe is a completely insignificant fraction. And of course, one could play Go on any $N \times N$ board. The number of actions would go like N^2, while the number of states would go like 3^{N^2}.

[2]There are 361 grid points, and each grid point has either a black stone, a white stone, or no stone on it.

The fact that it is seemingly easy to solve some problems (such as answering the question of what the product of two matrices is), while it seems much harder to answer others (such as answering whether a Diophantine equation has a solution) raises the question of whether some problems are simply harder than others. While we have no definitive answer as to whether this is the case, complexity theory characterizes problems into different complexity classes. The first (simplest class) is called P (for polynomial time). These are problems (taken from an infinite family of problems, such as matrix multiplication of two $N \times N$ matrices for any $N \in \mathbb{Z}_{\geq 0}$) whose solution time grows at most as a polynomial in N when computing them on an idealized computer called a Turing machine. The next class is called NP (for nondeterministic polynomial). For problems in this class, one can verify a proposed solution in polynomial time, with no statement as to how long it took to find the proposed solution or how it was even found in the first place. Since every problem in the class P can be solved in polynomial time, a proposed solution can of course also be verified in polynomial time, meaning $P \subseteq NP$. The name "NP" derives from the following. Let us envision the combinatorially large state space as a search tree. At each state, I can take N actions. This means that after k steps, I have N^k different states. If I have brute-forced the problem by searching the entire tree and found a solution, there exists a sequence of k_* actions, $(a_1, a_2, \ldots, a_{k_*})$, that takes me from my starting point to a solution. If I had a crystal ball (i.e., if I could act non-deterministically) and I could choose the right action at each level of the tree, I could solve the problem in k_* steps; or put differently, by working backwards from the solution to the starting position, I traverse k_* states. This means one can verify the positive answers in polynomial time (in k_* moves), which is the definition of NP that I prefer.

The question of whether there exist problems in NP that are not in P is the P vs. NP problem. If so, there would be problems (those in NP − P) that are genuinely harder than others (those in P) since, by definition, the former could not be solved in polynomial time. It is an open problem whether P = NP, but most people (including myself) believe that P ≠ NP. So, some problems are indeed harder than others [3]. In fact, there is a multi-trillion dollar bet against P = NP since internet security is based on the impossibility of factorizing a large integer into its primes — this is how the RSA encryption algorithm works. The P vs. NP problem is also a millennium prize problem, so if you resolve it either way you get $1 million. Hereinafter, we will assume that P ≠ NP.

Of course, we do not know all NP problems, which makes the following quite surprising: Cook [4] and Levin [5] proved that there exist NP problems that are as hard as can be in their class: they showed that there exists a problem in NP called SAT, which is as hard as any other problem in NP, in the sense that any other problem in NP can be reduced to this problem in polynomial time. Hence, if one could solve the SAT problem in polynomial time, one could first reduce some other NP problem one is interested in to SAT in polynomial time, and then solve the corresponding SAT problem in polynomial time. This means that SAT needs to be universal enough to encompass all NP problems. SAT is not some weird, abstractly constructed thing but quite easy to state: you are given a Boolean expression, meaning an expression in variables that can be either true or false, which are combined using Boolean operations such as AND, OR, and NOT, and you have to decide whether there exists an assignment of true/false to the variables such that the entire expression is SATisfied, i.e., true.

The SAT problem has all the ingredients we discussed above. Given a SAT problem with N Boolean variables, one has N actions (flipping the assignment of any of them from true to false, or vice versa), but there are 2^N different assignments — an exponentially large search space. The best known strategy to solve this problem is to try the exponentially many combinations until one works. However, if there was a fast (meaning polynomial time) algorithm to solve this problem, we would know that $\mathrm{P} = \mathrm{NP}$ since we could take any problem in NP, turn it into a SAT problem in polynomial time, and then solve the SAT problem in polynomial time. This means that SAT most likely cannot be solved quickly. Problems in NP that have the property that any NP problem can be reduced to them in polynomial time are called NP complete.

NP complete problems are by no means the hardest problems. There exist problems that cannot even be verified in polynomial time. This leads to a whole hierarchy of complexity classes:

$$\mathrm{P} \subseteq \mathrm{NP} \subseteq \mathrm{PSPACE} \subseteq \mathrm{EXPSPACE} \subseteq \mathrm{DECIDABLE}\,.$$

Like P and NP, these classes are also believed to be all different. Problems that are even harder than NP complete problems are again not weird or artificial, but occur frequently, for example in computational algebraic geometry. In this subject, one often needs to find a Groebner basis to solve certain divisibility problems, and finding Groebner bases is EXPSPACE complete [6]. If you wanted to compute the cohomology dimensions

$h^\bullet(X, V)$ of some vector bundle V over some manifold X (using Groebner bases) and someone gave you a proposed answer, there is no (known) way of verifying the proposal in polynomial time.

The last complexity class we mentioned, DECIDABLE, simply means that the problem can be decided without making any assumption about the time it would take to verify it. Again, most surprisingly, there are problems that are even harder than that. There are problems that do not have a solution. Such problems simply cannot be decided at all. Again, this sounds crazy, but it has been mathematically proven. The simplest such problem is the HALTING problem, in which you are given some code as input and should decide whether the code will eventually halt or run forever. Another undecidable problem, known as Hilbert's 10th problem and proven to be undecidable by Matiyasevich in 1970 [7], asks whether any given Diophantine equation has a solution. The difference between these problems and the NP problems like SAT mentioned above is that for the latter, the search space is bounded: given a Boolean expression in N variables, I can at least try all 2^N states and know for sure after I tried all of them whether one of them makes the Boolean expression true. For the halting problem, if I was simply running the code, at any given moment in time, I would not know whether I should continue waiting or whether it will just not halt. For quadratic Diophantine equations, there exists an upper bound for the maximum value one needs to try for all the variables in the polynomial. If one has not found a solution until then, the Diophantine equation does not have one. This bound goes away for higher degrees, and quartic Diophantine equations are in general undecidable, meaning that no matter how large the values one tried for all the variables, it could be that there exists a solution if one tried even larger ones.

7.2.2 *Introduction to reinforcement learning*

As mentioned in the beginning of the lecture, RL is a great way to navigate a huge (discrete) search space.[3] In a nutshell, instead of asking a crystal ball which action to take next to get to the solution of your task, you ask a neural network (NN). It is thus a promising tool to solve hard problems, such as problems in NP − P.

[3]There are RL algorithms for continuous state and action spaces, but we have found that RL works great for combinatorially large, discrete state spaces with rather small discrete action spaces.

In more detail, RL is based on behavioural psychology. The algorithm starts from some input state, which has a collection of legal actions it can perform to transition to the next state. The algorithm picks one of these actions and transitions to the next state. The action is then evaluated against the target objective and either rewarded (positively reinforced) or punished (negatively reinforced). The algorithm is trained to maximize its long-term return, i.e., the accumulated rewards/punishments it receives. One is optimizing the long-term return instead of the short-term rewards to avoid getting trapped in a local minimum. Using the board game chess as an example, you would potentially want the algorithm to make a pawn sacrifice or some other gambit, where you surrender material in favour of gaining a better position on the board in the future.

RL comes with its own vocabulary, which we have already used here and there, since it is rather self-explanatory. Nevertheless, it does not hurt to be precise, which is why we will define the various relevant terms more formally now. This is helpful for both talking about RL algorithms and reading/using/extending existing libraries.

Definition (Environment). The environment $\mathscr{E}$ is the set of states $\mathscr{S}$ and actions $\mathscr{A}$ on these states, $\mathscr{E} = \{\mathscr{S}, \mathscr{A}\}$.

Definition (State space). State space is the set of possible "'configurations" $\mathscr{S} = \{s\}$. We are interested in large (potentially infinite) discrete state spaces.

Definition (Action space). Action space is the set of actions that transition between the different states in the environment, $\mathscr{A} = \{a \mid a : \mathscr{S} \to \mathscr{S}\}$. We are interested in not too large discrete action spaces. Note that some actions might be impossible/illegal in some states. Actions can be stochastic, meaning that taking the same action will lead to different states (for example, an RL algorithm controlling a drone might issue a command to move it right, but a gush of wind might result in the drone moving left); however, we only consider deterministic action spaces.

Definition (Terminal states). The space of terminal states $\mathscr{T} \subset \mathscr{S}$ is the set of states for which no legal action is possible, i.e., the search ends in a terminal state: $\mathscr{T} = \{t \in \mathscr{S} \mid a(t) = \emptyset\}$.

Definition (Episode). An episode E is a sequence of states and actions that ends in a terminal state: $E = [(s_1, a_1), (s_2, a_2), \ldots, (s_n, \emptyset)]$ with $s_i \in \mathscr{S} - \mathscr{T}$, $s_n \in \mathscr{T}$, $a_i \in \mathscr{A}$.

Definition (Policy). The policy π describes how the algorithm (sometimes called agent) selects an action given its current state, $\pi : \mathscr{S} \to \mathscr{A}$. There are deterministic policies (meaning that the state uniquely identifies the next action) and non-deterministic policies (meaning that multiple actions are possible). One would want some degree of non-determinism for the policy in order to allow for exploration of the state space; if one follows a deterministic policy from the beginning, one will always only take the same trajectory through state space. Especially at the beginning of training, the algorithm should explore the environment and then eventually transition to exploiting what it has learned. Knowing when to make the transition from exploring to exploiting can be quite difficult and is known as the exploration versus exploitation problem.

Definition (Reward). The reward $r \in \mathbb{R}$ is the "feedback" given to the agent for following policy π, i.e., for performing action a in state s. We can thus think of r as the map $r : \mathscr{S} \times \mathscr{A} \to \mathbb{R}$. Often in RL, one faces the problem of sparse rewards. In the example of Go, one needs to finish an episode, i.e., play until one has reached a terminal state where the game is won or lost, in order to judge the quality of an action. It is not possible to just look at the board after an action and decide whether this was a winning or losing move (if it was, Go would not be much of a challenge for computers). This is to say, if there is an exponentially large search space with only one solution (one terminal state), the agent will only receive rewards/feedback and hence even learn the objective of the game once it has encountered this state in its random exploration of the environment. In such setups, RL should not be expected to perform better than a random search. In other instances, rewards can be non-sparse, which helps guide the algorithm towards the objective. For example, when trying to find the zeros of a Diophantine equation, one might give rewards after every action, for example a punishment that is proportional to the distance of the Diophantine equation from being zero.

Definition (Return). The return G is the accumulated reward (with punishments being negative rewards) from the current position onward, $G = \sum_{t=0}^{\infty} \gamma^t r_t$. In this formula, $0 < \gamma \leq 1$ is the so-called "discount factor". It discounts rewards in the distant future (as $t \to \infty$). If there is a big reward for solving the problem, a discount factor will incentivize the agent to solve a problem as quickly as possible.

Definition (State value function). The state value function $v(s)$ is the return G from the current state s_t onward, which can be expected for following policy π, $v(s) = \mathbb{E}(G \mid s = s_t)$.

Definition (Action value function). The action value function $q(s,a)$ is the expected return G for choosing action a_t in state s_t, $q(s,a) = \mathbb{E}(G \mid s = s_t, a = a_t)$.

This concludes the RL-specific vocabulary we want to introduce. Much more details can be found in Ref. [8]. So far, we have not mentioned NNs at all. We can think of RL as solving a Markov decision process (MDP). The reason why it is called a "decision process" is fairly obvious: the qualifier "Markov" refers to the fact that in deciding the next action, all that matters is the current state and not the trajectory through state space that got you there. The difficulty in solving the MDP is that the policy, the state value function, and the action value function are all complicated, unknown, interdependent functions. This is where NNs enter the game: they can be used to learn/approximate these functions. By optimizing the return, the quality of these function approximations is improved over time. Not all RL algorithms use NNs to approximate all these functions, but the most successful ones use NNs to approximate at least some.

7.2.3 *Summary*

Computational complexity provides means to classify problems according to how hard they are. Assuming all complexity classes are distinct (with the most prominent assumption implying $P \neq NP$), some problems cannot be solved quickly. For some problems in NP, it might be that we simply have not yet found an efficient way of solving them. For example, deciding whether an integer is prime can be done in polynomial time has only been discovered by Agrawal, Kayal, and Saxena in 2004 [9]. In contrast, any NP complete problem likely cannot be solved in polynomial time since otherwise $P = NP$.

Before blindly applying RL, or any ML algorithm, to a problem, I would recommend understanding what the complexity of the problem is (ideally even the average case complexity and whether there exist polynomial time approximation schemes) and whether there exist dedicated algorithms for solving these kinds of problems: when there are, I would definitely try these first. If there are not any (or if the best-known algorithm for the problem uses RL, as in the case of Go), a model-independent approach such as RL can be tried.

While RL algorithms are (rather) model independent, problems that are difficult because there is a huge space to search through and the best-known algorithm consists of trying all possibilities until one finds one that works make a great target for RL. While RL is also simply an algorithm that runs on a computer and thus cannot beat computational complexity (or even decide undecidable problems), it might still be able to recognize patterns or identify heuristics that can be used to obtain a good policy for solving a problem. This has been demonstrated impressively by DeepMind with their AlphaGo [1,2] and AlphaZero [10] algorithms. Ultimately, we do not know whether these are good Go players — after all, they only explore a completely insignificant fraction of the total search space of 10^{170} possible Go boards — all we know is that they are immeasurably better than any human, which might be good enough to make progress on a given problem.

References

[1] D. Silver, A. Huang, C. J. Maddison, A. Guez, L. Sifre, G. Van Den Driessche, J. Schrittwieser, I. Antonoglou, V. Panneershelvam, M. Lanctot, *et al.*, "Mastering the game of Go with deep neural networks and tree search," *Nature* **529**, no. 7587, 484–489, 2016.

[2] D. Silver, J. Schrittwieser, K. Simonyan, I. Antonoglou, A. Huang, A. Guez, T. Hubert, L. Baker, M. Lai, A. Bolton, Y. Chen, T. Lillicrap, F. Hui, L. Sifre, G. van den Driessche, T. Graepel, and D. Hassabis, "Mastering the game of Go without human knowledge," *Nature* **550**, 354–359, 2017.

[3] S. Aaronson, "$P \stackrel{?}{=} NP$," in *Electronic Colloquium on Computational Complexity (ECCC)*, Vol. 24, p. 4, 2017.

[4] S. A. Cook, "The complexity of theorem-proving procedures," in *STOC'71*, pp. 151–158. Association for Computing Machinery New York, NY, USA, 1971.

[5] L. Levin, "Universal sequential search problems," *Probl. Peredachi Inf.* **9**, 115–116, 1973.

[6] K. Kühnle and E. W. Mayr, "Exponential space computation of gröbner bases," in *Proceedings of the 1996 International Symposium on Symbolic and Algebraic Computation, ISSAC'96*, pp. 63–71. Association for Computing Machinery New York, NY, USA, 1996.

[7] J. V. Matijasevic, "Enumerable sets are diophantine," *Sov. Math. Dokl.* **11**, 354–358, 1970.

[8] F. Ruehle, "Data science applications to string theory," *Phys. Rept.* **839**, 1–117, 2020.

[9] M. Agrawal, N. Kayal, and N. Saxena, "Primes is in p," *Ann. Math.* **160**, 781–793, 2004.

[10] D. Silver, T. Hubert, J. Schrittwieser, I. Antonoglou, M. Lai, A. Guez, M. Lanctot, L. Sifre, D. Kumaran, T. Graepel, T. Lillicrap, K. Simonyan, and D. Hassabis, "A general reinforcement learning algorithm that masters chess, shogi, and go through self-play," *Science* **362**, no. 6419, 1140–1144, 2018. https://www.science.org/doi/pdf/10.1126/science.aar6404.

CHAPTER 8

REINFORCEMENT LEARNING FOR KNOT THEORY: LEARNING THE UNKNOT AND SLICENESS

FABIAN RUEHLE

8.1 Introduction

In the third lecture, we discuss applying what we have learnt so far to problems in knot theory. Knot theory appears in many guises in physics (connections to Chern–Simons theory [1], the knots-quiver correspondence [2], monodromies [3], etc.), mathematics (representation theory and low-dimensional topology), biology (protein folding), and many other fields. But even if the reader does not particularly care about this (beautiful) subject, they can still read the chapter as an illustration of concepts and lessons that transcend knot theory and can be applied to obtain rigorous results for discrete problems.

The paradigm is to obtain provably correct results for problems in discrete mathematics from reinforcement learning (RL). As explained in Section 1.1, neural networks (NNs) are black boxes in the sense that it is very difficult to interpret what the millions or billions of parameters are actually doing and their role in expressing the function that the NN learnt. For this reason, obtaining rigorous results from stochastic methods such as NNs follows a different route: given a binary decision problem of the form "Does this object have some property X", we ask the algorithm to return a truth certificate along with the answer if the answer is yes. We do not make any assumptions of what the algorithm does if the answer is no; in practice, it would simply not terminate.

In the context of RL, this means that we record the states and actions that an agent takes until it reaches a terminal state that solves the problem. Upon reaching this state, a human can then look at the states and actions and verify that they do indeed solve the problem — that is, the truth certificate. If the algorithm does not figure out a solution to the problem, it will not reach a terminal state and just keep looking. This type of setup is particularly useful if there are (strong) sufficient but not necessary criteria to detect the no instances. In practice, this typically means computing mathematical invariants that reveal obstructions to property X. The only failure mode is that the object might actually have property X, but the algorithm did not find a terminal state that would prove it. Hence, the algorithm can prove in a verifiable way that an object has some property X, but it can never prove that an object does not have this property.

8.2 Lectures 2 and 3

8.2.1 *Introduction to knot theory*

We are all familiar with knots, from shoelaces, ties, and entangled headphones. A mathematical knot is similar to these knots, with the crucial difference that mathematical knots are always closed. A more mathematical definition is the following.

Definition (Knot/link). A knot (link) is an embedding of a single (multiple) circle(s) into a 3-sphere, $S^1 \hookrightarrow S^3$, up to ambient isotopy.

We wrote "up to ambient isotopy" since we only want to distinguish knots that are genuinely different, meaning they cannot be deformed continuously into one another. Also, instead of S^3, we could have taken $\mathbb{R}^3$, but for the problems we will discuss, in particular the smooth Poincaré conjecture in 4D, we prefer thinking about the ambient space as S^3.

Given this (abstract) definition, there are multiple ways of actually representing the knot. Most of the representations project the knot from 3D to a 2D plane and then keep track of which strand goes over and which goes under the other in the projection. These points are the crossings of the knot, and they are enough to specify the knot, so all knot descriptions capture this data in one way or another. We want to introduce three different representations of a knot, known as the braid representation, the planar diagram code, and the grid diagram, respectively. There are many more, such as Dowker–Thistlethwaite codes, Gauss codes, and petal projections.

Braid representation

Braids can be studied in group theory independent of knots. We call the (Artin) braid group on n strands B_n. The group has $(n-1)$ generators σ_i, $i = 1, \ldots, n-1$, which move strand i over strand $i+1$; their inverses σ_i^{-1} move strand i under strand $i+1$. This action is associative, the identity is given by not moving any strand over/under an adjacent strand, and composition of two elements σ_i and σ_j simply means performing the two braiding actions consecutively. In general, braid groups are non-commutative. However, two generators $\sigma_i^{\pm 1}$ and $\sigma_j^{\pm 1}$ do commute if $|i-j| > 1$ since they operate on independent strands; see Figure 8.1 (left). Moreover, one can convince oneself (e.g., by drawing the actions as in Figure 8.1 on the right) that $\sigma_i \sigma_{i+1} \sigma_i = \sigma_{i+1} \sigma_i \sigma_{i+1}$ for $i = 1, \ldots, n-2$. So, in summary, the braid group B_n has the presentation

$$B_n = \langle \sigma_1, \sigma_2, \ldots \sigma_{n-1} \mid \sigma_i \sigma_{i+1} \sigma_i = \sigma_{i+1} \sigma_i \sigma_{i+1}, \sigma_i \sigma_j = \sigma_j \sigma_i \rangle,$$

where we have given the constraints on the indices above.

The relation between knots and braids is obtained from representing a knot as the closure of the braid; see Figure 8.2 (left). This means that the start of the first strand is identified with the end of the first strand, and likewise for all other strands, i.e., the braid is the object obtained from cutting open the knot. It is far from obvious that every knot can be represented as the closure of a braid; this result was proven by Alexander [5].

When describing knots as closures of braids, we need to take care of more equivalences. First, since the endpoints of the braid are identified, one can shuffle generators from one end to another. In other words, the braid words w and sws^{-1} describe the same knot, where s is any braid

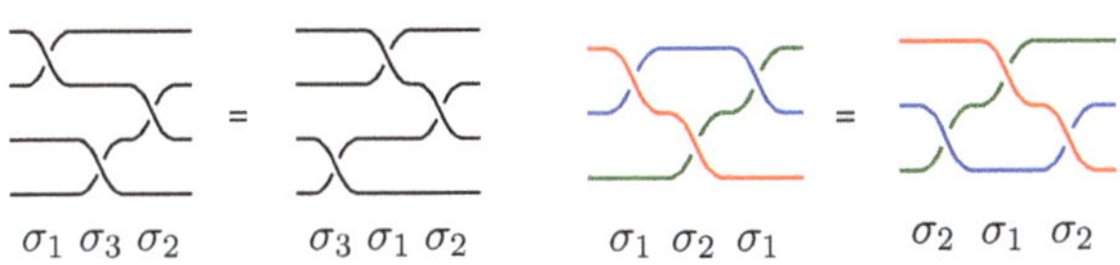

Fig. 8.1. Equivalence relations on a braid. Left: Operators commute if they operate on different strands, i.e., if $|i-j| > 1$. Right: Illustration of the relation $\sigma_i \sigma_{i+1} \sigma_i = \sigma_{i+1} \sigma_i \sigma_{i+1}$. Note how the strand crossing and their final positions stays the same.

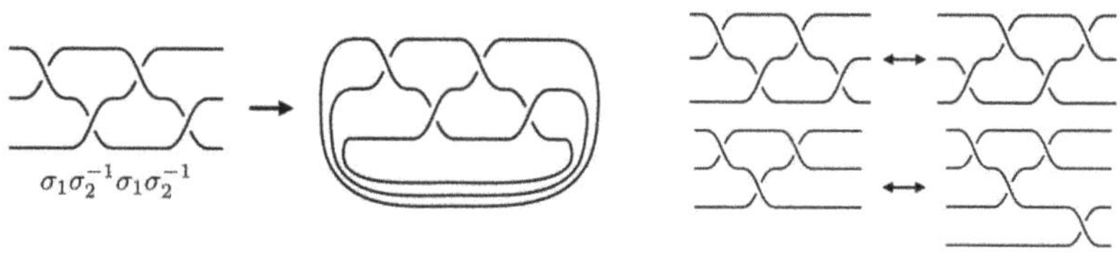

Fig. 8.2. Left: Knots can be described as the closure of a braid. Right: The two Markov moves: conjugation (top) and stabilization (bottom).
Source: Figure taken from Ref. [4].

word; see Figure 8.2 (top right). Second, one can use ambient isotopy to introduce a trivial "sling" in any strand. At the level of some braid word $w \in B_n$, this corresponds to an element $w' = w\sigma_n$ with $w' \in B_{n+1}$; see Figure 8.2 (bottom right). These two manipulations are called Markov moves. Finally, one should note that braid representations might not be minimal in the sense that manipulating a knot such that it becomes a braid might require introducing more crossings than necessary in other representations or projections.

Planar diagram code

Planar diagram (PD) codes are another way of representing the crossing information of a knot. For an n-crossing knot, a PD code is an unordered list of n ordered 4-tuples of numbers, and each number $i \in \{1,2,\ldots,2n\}$, appears exactly twice. The ordered tuples are formed as follows. Given an orientation (i.e., a direction of traversal) of a knot projection, choose an arbitrary crossing as the first crossing. This crossing will have two strands cross, giving four segments. Focus on the incoming under strand, and label the segments before and after the crossing with 1 and 2. More generally, the index of a segment on a strand is always increased by 1 upon encountering a crossing (at the nth crossing, it is increased by 1 mod n). This gives the tuple $(1,\bullet,2,\bullet)$, where the empty positions $\bullet$ corresponding to the two segments of the over strand are not yet determined; we simply list the four labels of each segment participating in the crossing counterclockwise in the tuple. Upon continuing traversing the knot, we will encounter this crossing one more time (on the over strand), and then we will know what the labels for $\bullet$ should be.

We follow the under strand to the next crossing and use the label 2 for the incoming strand segment of this crossing and label the outgoing strand segment as 3. In the second tuple, we record again the four numbers

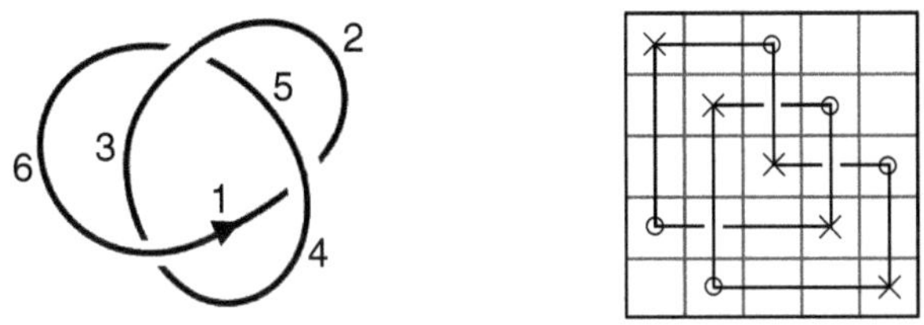

Fig. 8.3. Left: Strand labellings for the PD code in (8.1) of the trefoil. Right: Grid diagram for the trefoil.

labelling the strand segments, starting with the incoming under strand at position 1 and continuing counterclockwise. This means that if we were incoming again on an under strand, the tuple for the second crossing would be $(2,\bullet,3,\bullet)$. If we are on an over strand, it would be $(\bullet,2,\bullet,3)$ or $(\bullet,3,\bullet,2)$, depending on whether we first encounter segment 2 or segment 3 when proceeding counterclockwise from the incoming under strand. Upon encountering the crossing for a second time, we will know what labels to put for $\bullet$. We give the PD code of the trefoil in Figure 8.3 (left). Starting from the arrow, the process of building the PD code would be as follows:

$$
\begin{aligned}
&\{\} \to \{(1,\bullet,2,\bullet)\} \to \{(1,\bullet,2,\bullet),(\bullet,2,\bullet,3)\}\\
&\quad \to \{(1,\bullet,2,\bullet),(\bullet,2,\bullet,3),(3,\bullet,4,\bullet)\} \qquad (8.1)\\
&\quad \to \{(1,4,2,5),(\bullet,2,\bullet,3),(3,\bullet,4,\bullet)\} \to \{(1,4,2,5),(5,2,6,3),(3,\bullet,4,\bullet)\}\\
&\quad \to \{(1,4,2,5),(5,2,6,3),(3,6,4,1)\}.
\end{aligned}
$$

Grid diagrams

The third knot representation we want to discuss are grid diagrams. In contrast to the other two, these are a more diagrammatic way of representing a knot. As the name suggests, one starts with an $m \times m$ grid. In each row and each column, there is exactly one $\times$, one $\circ$, and $n-2$ blank grid cells. Now, you draw the (projection of the) knot by connecting the $\times$ to the $\circ$ in each column and the $\circ$ to the $\times$ in each row. By convention, whenever a horizontal segment intersects a vertical segment, it is assumed that the vertical segment passes over the horizontal segment. In practice, the knot is traversed by starting with the $\times$ in the leftmost column, say, moving up or down until we hit the $\circ$, moving right until we hit an $\times$ again, moving up and down to the next $\circ$, and so on. We give the grid diagram of the trefoil in Figure 8.3 on the right.

With this wealth of possible representations, one might wonder which representation to choose. In principle, it should not matter: each representation characterizes a knot uniquely, and there exist algorithms to transform a knot from one representation to another.

In practice, one should think about the following for choosing a data representation (this is true in general, not just for knot theory). From a computer science point of view, one might ask which representation allows for the most efficient encoding, meaning which representation needs the least amount of memory and processing time. However, it could be that the theoretically most efficient encoding is not the best one for your ML algorithm of choice. A second question one should ask is whether the problem can be cast into a form that resembles some problem that is standard in the CS community. This will mean that plenty of prior research as well as optimized code libraries will be available. Third, from a mathematical point of view, one should ask whether some presentation has a beneficial property that others do not. In the following, we illustrate these three points using simple examples from knot theory.

Efficiency of the data representation

While it is very common to work with knot projections, it might be more efficient to work with the three-dimensional object. Essentially, upon projecting, one produces many crossings, and as we saw above and will make more precise later, knots are identified by their crossings. This was formalized in Ref. [6], where the authors argued that some knot or link invariants can be computed more efficiently in 3D rather than in 2D. But even among the different ways of representing a knot projection, some require more data than others. For example, a PD code uses four integers per crossing, while a Gauss code would only use one. Braid words also use only one integer ($\pm i$, encoding in which $\sigma_i^{\pm 1}$ was used), but they have the added complication that they might require more crossings than the minimum number that can be used in Gauss or PD codes.

Apart from these considerations, one should ask whether the algorithm in use can actually make use of the more efficient encoding. Simply because there exists an efficient algorithm does not mean that the machine will also learn it, or it could require a different NN architecture. Moreover,

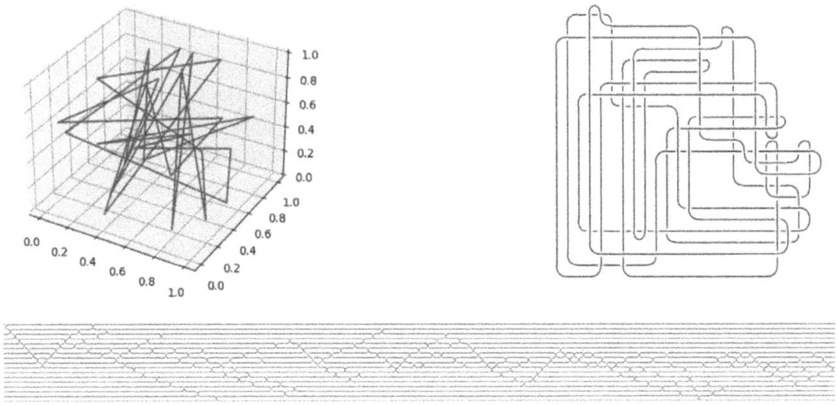

Fig. 8.4. A knot generated by a random walk in 3D (top left), a knot projection from which the PD code was calculated (top right), and the braid configuration (bottom).

providing in principle redundant features can improve NN learning in some cases and impair it in others. Finding a good set of features is called feature engineering; see Ref. [7] for more details.

To illustrate this in a simple example, we generated random knots via a random walk in 3D. These random knots were then projected to 2D along a random direction, and we computed various representations of the knot, which we used to compute a knot invariant, called the hyperbolic knot volume. The hyperbolic knot volume is conjectured to be computable more efficiently in 3D in Ref. [6]. We illustrate the procedure in Figure 8.4, where we give the knot obtained from a random walk, a projection of the knot used to compute the PD code, and the corresponding braid representation. The 3D knot is represented by 25 points, each with three coordinates (so 75 numbers), the PD code has 87 crossings (so 348 numbers), and the braid representation has 185 crossings (so 185 numbers).

We set up a supervised learning task where we generate 10,000 knots, compute their hyperbolic volume[1] using `snappy` [8] and split off 10% as a test set. We train a simple feed-forward NN with three hidden layers of 64 nodes each and ReLU activation function (including at the output layer since the volume is non-negative) for 100 epochs using an ADAM optimizer and record the mean square error. We find that the predictions

[1]At a low number of crossing, the majority of random prime knots are hyperbolic; we perform rejection sampling and resample if a knot is not hyperbolic.

based on PD codes are the best, followed by braid words and then by the 3D representation. This means that the least efficient feature representation gives the best results, and the most efficient representation, which is also the theoretically superior one for this task, shows the worst performance. A reason could be that the architecture is suboptimal for the case at hand. Fully connected feed-forward networks are invariant under permutations, while the order of the 3D coordinates or of the braid generator codes matter to uniquely identify the knot. The order of each tuple in the PD code also matters, but the tuples themselves can be permuted.[2]

Architecture for data representations

The different encodings of a knot might inform the NN architecture one should use. For example, when using braid *words*, one might use an architecture from natural language processing, such as the transformer. Grid diagrams, on the other hand, might benefit from architectures used in computer vision such as CNNs or graph NNs.

In Ref. [4], we tested different representations for the unknot recognition problem. The problem is a binary decision problem that asks whether a given knot representation is the unknot, i.e., the unique knot with zero crossings. We used a reformer [9] and a vanilla feed-forward architecture, which is a transformer with some modifications to make it less resource-intensive. This indeed produced the best results when using braid words as inputs, with Gauss or DT codes trailing behind in performance. However, the pattern also continued when using a feed-forward NN.

Mathematical properties of the data representation

In some cases, one representation might have mathematical properties that are beneficial over those of the other. Sticking with the unknot example, it is known that there exist "hard unknots", which require introducing more crossings before the knot can be simplified to the unknot [10, 11]. Given a knot that has this property in DT code representation, say, it is not clear that this property is preserved when mapping the knot to a braid word (which might need more than the minimal number of crossings anyway). For grid diagrams which are simplified using the so-called

[2]One can get good performance even if the architecture has the "wrong symmetries"; Google's AlphaZero uses convolutional layers even though chess is not translationally equivariant.

Dynnikov moves, it has been shown that the number of crossings never increases when simplifying the knot to the unknot [12]. For this reason, Kauffman *et al.* [13] used this setup for an unknot recognition task. In contrast, we used RL with a reformer or feed-forward architecture and braid word representations. Since RL is trained to optimize the long-term return instead of the short-term reward, the algorithm can learn to make the knot temporarily more complicated if it then gets a big reward for removing all crossings in the end. This shows that mathematical properties of the representation should inform the algorithm one chooses for the task.

8.2.2 *Knot theory as an RL problem*

As discussed at the end of Section 7.1, we are interested in obtaining provable results for binary decision problems in discrete mathematics. More precisely, we set up an algorithm which is supposed to halt on the "yes" instances of the decision problem and produce a truth certificate that can be checked/interpreted by a human, while it is allowed to run forever and/or not produce a certificate for the "no" instances. We will illustrate the use of RL for the following decision problems:

Decision Problem (UNKNOT): Does a given knot representation represent the unknot?

Decision Problem (RIBBON): Is a given knot a ribbon knot?

We already described what we mean by the unknot above. For ribbon knots, there are different ways of introducing them. While the Morse-theoretic picture might be more intuitive, the following definition will be more useful for our purposes. Note that a knot embedded in 3D bounds an (in general self-intersecting) 2D disk in S^3; see Figure 8.5 for an example.

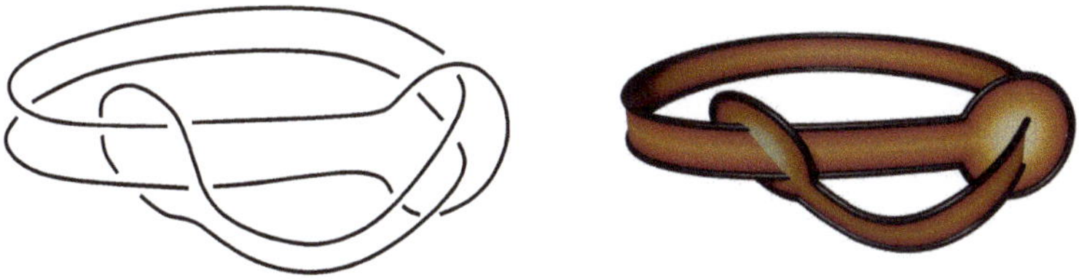

Fig. 8.5. The square knot $3_1\#3_1^*$ is ribbon. It bounds a self-intersecting disk in 3D with two ribbon singularities. In 4D, the disk does not self-intersect (we color-code the fourth dimension into which we can pull the disk to avoid self-intersection in the right figure).
Source: Figure taken from Ref. [14].

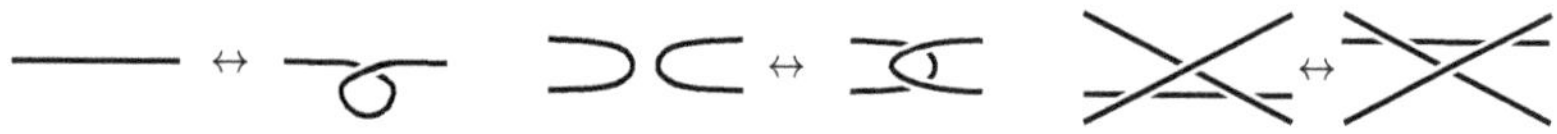

Fig. 8.6. The three Reidemiester moves: twist/untwist (Reidemeister) (left), poke/unpoke (middle), and slide (right).

If all self-intersections are ribbon singularities, which are self-intersections that correspond to cutting a slit into the disk and pulling another part of the disk through the slit, the knot is called a ribbon knot. Now, we consider the 4-ball B_4 whose boundary is the S^3 in which the knot lives. If the knot bounds a (smooth) disk in B_4, it is called (smooth) slice. As it turns out, every ribbon knot is a slice. Away from the ribbon singularities, the knot already bounds a disk in S^3. At the ribbon singularities in S^3, one can pull the self-intersecting part into the fourth dimension to obtain a smooth disk in 4D. The converse, i.e., whether every slice knot is ribbon, is an open problem known as the slice-ribbon conjecture [15]. Nevertheless, we will be not too careful in distinguishing between them in this lecture. Our algorithm is set up to detect ribbon knots, which then includes all slice knots that are also ribbon.

Both the UNKNOT and RIBBON problems seemingly require continuously deforming the knot: in the case of the unknot, we are asking whether a knot can be disentangled, and for ribbon, we are asking for a smooth embedding. So, it is not *a priori* clear what the relation to discrete mathematics is. One problem we have glossed over so far and which is probably bothering you if and only if you are a mathematician is whether non-trivial knots exist at all? In the end, it could be that all "different" knots are simply different representations of the unknot. Surely enough, non-trivial knots exist, and the UNKNOT problem is actually non-trivial. To show the existence of non-trivial knots, Reidemeister proved [16] that it is possible to relate any two knot projections of the same knot with a sequence of moves called Reidemeister moves. There are three different types of Reidemeister moves, as seen in Figure 8.6, and all involve local manipulations at knot crossings.

8.2.3 *Example* 1: *Unknotting*

The UNKNOT decision problem is an interesting problem from a computational complexity point of view since both the yes and no instances can be verified in polynomial time (meaning it is in NP $\cap$ co-NP [17, 18]), and Lackenby announced an unpublished result that unknot recognition can be done in a time that scales as $\mathcal{O}(n^{c\log(n)})$ with the number of crossings, where c is a constant. Also, the UNKNOT recognition problem can be considered a much simpler version of the RIBBON problem (which requires unknot detection in the way we set it up).

Since we are interested in truth certificates for the yes instances of the UNKNOT problem, we set up the algorithm to find a sequence of Reidemeister moves that explicitly untangles the knot as much as possible. If the input was the unknot, all crossings can be removed, and we will be left with only a simple circle without crossings in the end. Formulated this way, the UNKNOT problem is a specific case of the more general problem of reducing a representation of a knot to an equivalent representation with the minimal number of crossings.

In our paper [4], we decided to work with braid words and tools from large language models (transformers [19] and reformers [9]). This was motivated by the tremendous success of transformers in this area. For braids, the three Reidemeister moves are mapped onto the braid relations and Markov moves; see Figures 8.1 and 8.2. The Reidemeister twist becomes the Markov stabilization move, the Reidemeister poke corresponds to the Markov move of conjugation, and the Reidemeister slide corresponds to the second braid relation. In terms of braid words, the unknot corresponds to the empty word, and the UNKNOT decision problem reduces to a shortest word problem in the closure of the braid group. Phrased this way, the UNKNOT decision problem is a shortest word problem in a non-Abelian group. The most general word problem reads as follows:

Decision Problem (WORD): Given two words w_1 and w_2 in a non-Abelian group G, decide whether they correspond to the same group element $g \in G$.

It has been shown that this problem is undecidable [20]. However, the complexity of the UNKNOT problem is much better behaved. For braids, the word problem is actually solvable in P. The reason why UNKNOT is not known to be in P so far, even though the braid word problem is, is that inequivalent braids can represent the same knot.

As a first step, we need to think about how to generate our dataset. The problem in this case is of course that we cannot recognize the unknot; therefore, generating knots randomly to obtain some unknots is unfeasible (moreover, if you simply generate knots randomly, they will have a very low probability of being unknotted unless you know how to skew your prior). For that reason, we simply start from the solution, i.e., the empty braid word, and perform random (with a flat prior) Markov moves and braid relations until we obtain a knot with a predefined number of crossings. By construction, this knot will be the unknot.

Before applying RL to the problem, we check the performance of different NN architectures in a supervised binary classification task. We generate 10,000 knots, out of which 5,000 are the unknot and 5,000 are non-trivially knotted. There are many algorithms to construct random knots. Random knots will be, as mentioned above, typically not be unknots, which can also be checked by computing a knot invariant called the Jones polynomial. If the Jones polynomial is non-trivial, the knot is not the unknot. The converse is an open problem.

We want to make sure that the unknot and the non-trivial knots in our dataset are "as similar as possible". Otherwise, it could be that the NN picks up some signal in the sample bias that has nothing to do with the unknot problem but rather with the way the knots from the two different classes were generated. For that reason, we generate the non-trivial knots by starting with a random knot with a smaller number of crossings and then also perform random Markov moves and braid relations until we obtain a knot representation with the same number of crossings as the representations we constructed of the unknot.

In our binary classification experiments, we find that both feed-forward NNs and transformers/reformers can predict the correct class with more than 95% accuracy for a wide range of crossings. While reformers performed slightly better, the gain in accuracy does not justify the longer training time of these more complicated architectures. Essentially, if the feed-forward NN already performs at 95% accuracy, there is not much room for improvement. The fact that the feed-forward NN performs so well is surprising since we do not use positional encoding, and so *a priori* it does not know about the ordering of braid generators in the input; however, whether a knot is trivial or not depends on the order of the generators. This indicates that the NN is learning a positional encoding, among other things.

Of course, this binary classification result does not help us achieve our goal of having provably correct predictions since it involves a black box NN. Nevertheless, the great performance is, in my opinion, a good hint that there is some pattern or map in the dataset that the NN can learn/extract, and hence there is a very good chance that the NN in the RL will be able to learn good policy, state, and action-value functions. In the worst case, one could set up the most simple-minded policy NN, where we build a search tree by performing each possible action on the input knot and query our previously trained binary classifier to see after which action it is most sure that the resulting knot is the unknot. We then carry out this action (in a greedy setup, otherwise we could rank the actions according to the NN classifier output and draw the next action from this ranking according to some probability measure) and repeat until we reach a terminal state or give up.

But of course there are much more sophisticated RL algorithms available. Let us discuss how we setup the RL "game":

State space: The states are just the braid words, where we represent a generator $\sigma_i^{\pm 1}$ simply by $\pm i$ rather than tokenizing and using one-hot encodings. One subtlety is that actions can change the length of the braid word while the input of the NN stays constant (at least for the feed-forward case), which is why we zero-pad all braid words to a fixed length. Since there exist hard unknots whose length needs to be increased, we zero-pad braid words of length N with N additional zeros to allow space for length-increasing moves.

Action space: The actions are the two braid relations and the two Markov moves, plus a simplification move that removes consecutive inverses. These can, in principle, be carried out at any position in the braid. In practice, some actions are illegal. For example, the braid relations cannot be performed at any position in the braid. This makes the action space quite large. We found that a much better performance can be reached by a modified action space, which we will describe later.

Terminal states: The terminal state is the empty braid word corresponding to the unknot. Moreover, we set a timeout of 300 steps. If the agent has not reached the terminal state after 300 steps, the episode is ended. This is necessary since the terminal state cannot be reached if the input is not the unknot. In such cases, the algorithm will idle and simply perform an action that does not incur a negative reward.

Rewards: Since we do not know *a priori* the shortest possible braid word, we cannot reward the agent with the distance to the optimal solution. Instead, we only punish the agent, where the punishment is the negative of the length of the braid word. Another approach would be to reward or punish the agent by the difference in length of the braid word after carrying out the action. We also experimented with punishing illegal actions.

Instead of punishing illegal actions, one can simply ignore them and re-query the policy NN (for non-greedy setups) or mask out all illegal actions.

Policy and value NNs: For the policy NN, we use a feed-forward NN with $2N$ input nodes (the braid word), three hidden layers with 200 nodes each, and an ReLU activation function. The output layer has $N+6$ nodes, one for each action. The value NN has the same architecture, except that it has a single output node that indicates the value.

As mentioned above, we obtained much better results by making the action space smaller. For a braid word of length N, there are N actions that specify where the first braid relation should be used, N actions where the second braid relation should be used, $2N$ possible conjugations (or rather, this should be the number of braid generators plus their inverses), one stabilization move, one destabilization move, and N simplification moves that remove consecutive inverses at position N, for a total of $5N+2$ actions. In addition to being a large action space, it also contains many illegal actions. To improve on this situation, we implement a pointer that points to a specific position in the braid. We then have only two actions for the two braid relations, which will be applied at the first legal position encountered to the right of the pointer (and wrapping around at the end of the braid word). For the Markov moves, we still have N and one move, respectively. We modify the simplification move to eliminate consecutive inverses everywhere in the braid word, so there is only one such action. Finally, we add two actions that moves the pointer left or right. While one direction would have been sufficient theoretically (as we wrap around at the end of the braid word), the NN would potentially have to perform $\mathcal{O}(2N)$ pointer moves before it can carry out the move it is interested in. This requires the NN to capture very long-range correlations, especially for long braid words. Moreover, after performing an action, the pointer is automatically advanced one position to the right. In total, this leads to $N+6$ actions.

For our RL runs, we try different RL algorithms, as implemented in the `chainer-rl` library [21]. We train different agents with braid words of lengths 12, 24, 36, 72, and 96. Given a knot of length 50, say, one could then use the RL agent that was trained on length-72 knots (by zero-padding the 50 crossing knot or destabilizing until it has length 72) and run it until it has simplified the braid word to a word of length 36. Then, we can use the "smaller" RL agent that has been trained on braid words up to length 36 to further simplify the knot until the braid word length reaches 24, and so forth. The advantage of this is that the NNs for smaller inputs have a smaller number of parameters and can hence be queried faster. Moreover, if one uses the NN for the length-72 braid word all the way down to the unknot, the braid word would, in the end, essentially consist of padded zeros only, which might confuse the NN. In practice, we found that these issues did not arise in our case, and we simply used the same network all the way to the terminal state.

To evaluate the performance, we compare different RL algorithms to the performance of a random walker, which simply draws actions randomly with a flat prior. We run the agents on 1,000 non-trivial representations of the unknot and record for how many it reaches the terminal state. In that case, we record the actions as the truth certificate for the input knot being the unknot. We find that the performance of the random walker decreases exponentially with braid word length. The best performing algorithm is Trust Region Policy Optimization (TRPO), which reaches the terminal state for $>80\%$ of the test knots, across the range of crossings we tested (i.e., up to 96 crossings).

8.2.4 *Example 2: Ribbon and slice knots*

The RIBBON problem is relevant to the smooth Poincaré conjecture in 4D (SPC4), which asks whether every manifold that is homeomorphic to the standard 4-sphere[3] is also diffeomorphic to the 4-sphere. This is the last open conjecture of the Poincaré conjectures (with the 3D version famously proven by Perelman [22, 23]). The conjecture is very interesting since it is known to be true in dimensions 1, 2, 3, 5, and 6 (among others), but false in general. It is also related to other open conjectures, for example in representation theory.

To make the connection with ribbon knots, we use the work of Manolescu and Piccirillo [24]. They constructed a pair of knots which are related in a certain way (by zero-surgery to be precise). If one of the knots is slice and the other is not, one can use them to construct an exotic 4-sphere, and SPC4 would be false by counterexample. There are knot invariants that can be used to show that a knot is not ribbon or slice, for example the signature, Rasmussen invariant [25], tau invariant [26], the Fox–Milnor criterion [27], or the Herald–Kirk–Livingston criterion [28].

[3]The standard sphere is obtained by gluing two disks along their boundary with the identity map.

However, they do not tell you when a knot is ribbon (for example, if the tau invariant is larger than zero, the knot is not slice, but if it is zero, the knot could or could not be slice; this is because the tau invariant is a lower bound for the slice genus, which is zero for a slice knot, but we do not know whether the bound is saturated). Hence, we would want to find a pair of knots related by zero-surgery, where one knot has a slice obstruction and the other one does not. We then run the RL algorithm on the non-obstructed knot, hoping to find that it is a ribbon knot. The algorithm would give us a truth certificate, so we could prove that it is indeed a ribbon knot and SPC4 is false.

In order to check that a knot is ribbon, we use link concordances. Link concordance is an equivalence relation, under which two links L_0 and L_1 are called concordant if there exists an embedding

$$f : L_0 \times [0,1] \to L_1 \times [0,1]$$

such that

$$f(L_0 \times \{0\}) = L_0 \times \{0\} \quad \text{and} \quad f(L_0 \times \{1\}) = L_1 \times \{1\}.$$

By adding the binary operation of taking the connected sum of two links, $L_0 \# L_1$, we obtain a group called the concordance group.

Every ribbon knot is concordant to the unknot. This means that if we can add moves that perform ribbon concordances in addition to the simplification moves we discussed in Section 8.2.3, we should be able to reach the unknot if the starting knot is ribbon. The main new actions to capture link concordances are band additions. You can think of a band addition as follows: cut the link open at two positions and close the link up again by connecting the two cuts by two parallel strands. In principle, these strands need not be parallel; they are allowed to self-intersect with a ribbon singularity. In practice, this makes our state and action spaces complicated, so we restrict to concordances without ribbon singularities in the band. However, we do allow twisting the band multiple times before reattaching it at the second cut. In connecting the two cuts, we can route the band however we like. In particular, it is allowed to go over or under existing strands in the knot projection or wrap around them (although we do not allow for the latter possibility for the same reason why we do not allow for ribbon singularities). We present an example of a band plus simplifications for the simplest ribbon knot, the Stevedore knot *K6a*1, in Figure 8.7.

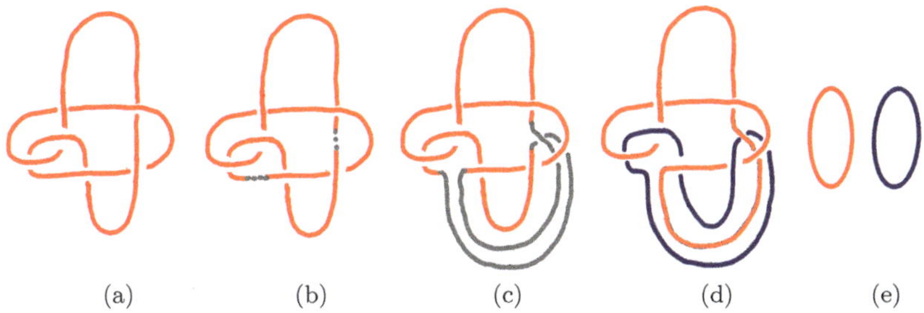

Fig. 8.7. We (a) start with the Stevedore knot, (b) cut it open at two positions (dotted grey lines) and (c) add a band with a twist (grey). (d) This gives a two-component link (red and blue). (e) Subsequent simplification moves lead to two unlinked unknots. This proves that the Stevedore knot is ribbon.

The band additions have to be chosen such that:

- a band only connects different parts of the same link component,
- bands are compatible with orientation, and
- band additions increase the number of link components by one.

We implemented band additions using the python package snappy [8]. For the knot simplification, we use snappy's simplify function, which uses heuristics instead of an NN, as in Section 8.2.3, to simplify a link.

In order to train the algorithm, we need to generate ribbon knots. In Ref. [14], we implemented two algorithms to generate them, which we called SYM and UNSYM. They are illustrated in Figure 8.8.

SYM. The SYM algorithm makes use of the fact that a knot projection that is mirror symmetric about an axis is ribbon [29]. To construct such knots, we construct a random braid word w in B_{2n}. This will form the left side of the knot. The mirror of w is simply w^{-1}. We can further allow for two crossings along the symmetry axis. For these we construct a braid word s which contains exactly $n-1$ non-consecutive[4] braid generators $\sigma_i^{\pm 1}$. We then form the braid word $r = wsw^{-1}$. If we now took the closure of the braid in the usual way, we would obtain a link that is just described by s. To see this, note that we could conjugate by $r \to w^{-1}rw = s$. This is not what we want, so we generate a knot from r by identifying strands 1 and 2, 3 and 4, etc., on both ends of the braid; this is why we constructed the braid word in B_{2n}. The procedure is illustrated at the top of Figure 8.8.

[4]This condition means that the two crossings are along the symmetry axis.

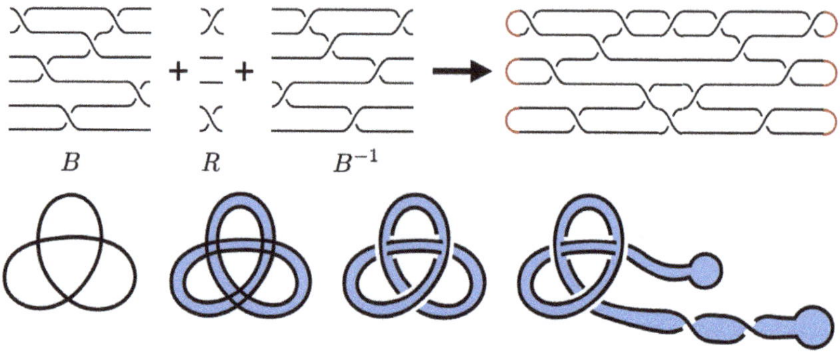

Fig. 8.8. Generating ribbon knots. Top: SYM algorithm. Bottom: UNSYM algorithm.
Source: Figure taken from Ref. [14].

UNSYM. The UNSYM algorithm constructs knots which bound, by construction, disks with at most ribbon singularities. We start with a random knot (constructed using snappy's random knot generator). We then double the knot such that it bounds a disk between the two knot components. At the crossing of the knots, we need to choose an intersection. We can either put one part of the disc in front of the other, put it behind the other, or let the disk self-intersect in a ribbon singularity. At each crossing, we choose one of these possibilities randomly with a flat prior. We illustrate the procedure at the bottom of Figure 8.8.

Next, let us discuss how we set up the RL algorithm.

State space. To describe a band, we need to keep track of where it starts, which strands it passes (and whether it passes over or under them), and where it ends. In terms of the knot projection, this means we choose a start strand and then decide over/under which adjacent strand we move next. Other options include twisting the band or reattaching it at an adjacent strand. We choose to represent the knot via a (modification) of its dual graph. In the dual graph, the vertices correspond to the "voids" in the graph, and two nodes are connected by an edge if the two void regions are separated by a strand in the knot projection. We represent the graph by its adjacency matrix. However, the dual graph does not capture all the information about the knot (for example, whether crossings were over- or under-crossings), which is why we choose an orientation for the knot and then decorate the edges by a sign indicating whether the strand goes over or under the next crossing. We also enumerate the crossings. A knot with N crossings has $N+1$ voids (for example, the unknot with zero crossings has a void inside the circle and a void outside the circle), so we represent the knot by an $(N+1) \times (N+1)$ matrix. While strictly not necessary to describe the link uniquely, we also add a "component matrix", which is simply the adjacency matrix of the dual graph of the link, but we enumerate the edges according to which link component they belong. The reason behind this is that legal band additions can always only connect a band between strands of the same link. This component matrix is another $(N+1) \times (N+1)$ matrix. To describe the band, we specify the order in which the band traverses the vertices of the dual graph. We describe the dual graph again by an adjacency matrix and label the order in which the vertices are visited by corresponding integers on the diagonal. For the off-diagonal entries, which encode the edges of the dual graph, we use ± 1 to indicate whether the band moved over or under the corresponding strand. Hence, the description of the band requires a third $(N+1) \times (N+1)$ matrix. Finally, we need to indicate how often the band was twisted and in which direction. For this, a single (signed) integer would be enough. However, since we already have three $(N+1) \times (N+1)$ matrices, we represent the twists by another $(N+1) \times (N+1)$ matrix, where each entry is the signed integer indicating the number of times the band has been twisted.

Action space. We model the band as a self-avoiding (random) walk. As explained above, the walk does not have to be self-avoiding on mathematical grounds; we could have ribbon singularities or have the band pass over/under itself. We avoid these since it would require a modification of the state space, i.e., how we describe the band. For a knot with $N+1$ dual graph vertices, we have $N+1$ actions that indicate where the band should start. After starting the band, there are $2(N+1)$ actions that describe which vertex we move to next; the factor of two is because we can move over or under a strand when going to the next dual graph vertex. Note that out of these $N+1$ actions, most will be illegal since we can only move to an adjacent vertex. Moreover, once a band is started, the start actions become illegal until the band has been attached and a new one can be started. We also have $(N+1)$ attach moves that indicate where to attach the band; again, most of these actions are illegal since a band needs to be adjacent to the vertex that describes where the band currently is. Furthermore, a band cannot be attached before it is started, and it can only be attached to the same link component on which it started. Once attached, we simplify the knot (using snappy's simplify function) and check whether the resulting link consists of unlinked unknots only. Lastly, there are two moves that will insert a positive or negative twist, respectively. We want to avoid that the agent twists the band back and forth (this would cancel out), so once the agent twisted in one direction, twisting in the other becomes an illegal action. If the knot provided to the unknot is not ribbon, the agent will idle and perform band additions that simply split off an unlinked unknot, i.e., leaves the link unaltered. We mask out all illegal actions.

Terminal states. Terminal states are the unknots. Moreover, since we impose that each vertex can only be visited once, the agent can become trapped with nowhere to move. In particular, if all vertices adjacent to the current vertex belong to link components that are different from the one the band was started on, no moves are left (except for twisting; since we do not want the agent to twist indefinitely in such a situation, we limit the maximum number of allowed twists and treat it as a hyperparameter). If the agent gets trapped or if the attached band is invalid, we end the episode, discard the band, and start over.

Rewards: As in the UNKNOT problem, we want to reach a link with zero crossings. Hence, we can punish the agent again by choosing a negative reward proportional to the number of crossings in the link. Note that the rewards can be rather sparse since attach moves are the only moves that can change the number of crossings.

Policy and value NNs: Since our state space consists of four $(N+1) \times (N+1)$ matrices (or "channels"), we decided to choose an architecture from computer vision. We tried ResNets [30] and simple LeNets [31]. The simple LeNets performed almost as good as the ResNets and can be trained and queried much faster, which is why we decided to use them in the end.

As an initial experiment, we phrased the ribbon problem again as a binary decision problem. Unlike the UNKNOT problem, we obtained much worse results, with around 65% accuracy. This indicates that the ribbon problem is much harder than the unknot problem. For the RL agents, we used the same chainer-rl agents, as for the unknot problem. We also modified the alpha zero search tree algorithm to work for single player games such as the ribbon problem. The problem with the alpha zero algorithm is that it performs a search over legal actions in building the search tree. Since some of the attaching moves are almost always legal, the agent performs them a lot. After each attach, we must check whether the game is won, i.e., whether the terminal state of the unknot was reached. This requires solving the unknotting problem after each attach move, which makes attach moves computationally very costly.[5] In the end, we did not have enough resources to train the algorithm at scale and cannot comment on its performance.

From inspecting rollouts of the trained RL agents, we noticed that they tend to prefer short bands and small numbers of twists. This led us to also look at a random walker (RW) that does not draw from a flat prior but prefers for example attach moves over other moves. This random walker is state independent, i.e., it decides actions randomly and not based on the current state (although illegal actions, which depend on the state, are masked out). In total, we have then the same probability for all $N + 1$ start moves, all $N + 1$ attach moves, etc., but allow different probabilities for each such type of action. To find the optimal probability distribution, we use a Bayesian optimizer [32] implemented in Ref. [33]. We choose this over other hyperparameter optimization frameworks since it shows good performance with only little simulations necessary, and each simulation is highly time-consuming because we want to optimize hyperparameters over a wide range of knots.

When comparing the different algorithms, we find that the A3C and TRPO algorithms work equally well (both with LeNets or ResNets). However, they perform on par with the Bayesian optimized RW. This remained true when we started from the Bayesian optimized probability distribution and used the policy NN to learn a correction (or state-dependence) for the individual action probabilities. Moreover, out of the knots with unknown ribbon status, both the RL agent and the Bayesian optimized RW recognized the same ones to be ribbon.

We also noticed a sample bias in our knot generation algorithms: the UNSYM knots are much simpler to solve than the SYM knots. The latter are as hard as a ribbon knot taken from snappy's database of all knots up to 14 crossings. In fact, essentially all UNSYM knots can be solved by either algorithm in a fraction of a second for up to 50 crossings and within a few seconds up to 100 crossings. In contrast, solving SYM knots up to 50 crossings can take up to 5 minutes on a 2021 Macbook Pro for some knots and even longer for 100 crossing knots. Most of the ribbon knots with less than 15 crossings are found to be ribbon in a fraction of a second, although some (previously unknown) knots with up to 14 crossings were shown to be ribbon after several core-weeks of runtime.

---

[5]Note that this is different from games like go or chess, where checking whether the game is won is not more expensive than finding legal actions.

Besides our generated set of SYM and UNSYM ribbon knots, we ran the Bayesian RW and the trained A3C algorithms on the set of potential counterexamples to SPC4 constructed by Manolescu and Piccirillo [34], some of the GST examples [35], and the alternating bounty knots from Owens-Swenton [36]. For the 3375 RBG links from the Manolescu–Piccirillo set, we managed to rule out all but two pairs of knots as potential counterexamples tp SPC4. For the GST knots and the bounty knots, the algorithm did not find ribbon knots. If you want to try the code yourself, it is available on my GitHub page [37].

8.2.5 *Summary*

We applied RL to open problems in mathematics. We discussed applications to knot theory, but the idea is applicable to problems in discrete mathematics more generally. In essence, both examples come down to solving a shortest-word problem in a non-Abelian group up to some equivalence relations.[6] The algorithm is set up to find a sequence of actions that reduces a given input to the empty word. If it succeeds, the sequence will present a truth certificate for the yes instances of a binary decision problem. We do not make any assumptions about the behaviour of the algorithm for the no instances. In our setup, the algorithm will not provide an answer because it is interrupted after a maximum number of steps (as in the unknot case) or does not halt at all (as in the ribbon case). In both examples, one can compute knot invariants that, if non-trivial, will unambiguously establish that a given input is a no instance; if the invariant is trivial, it could be a yes or no instance. In the case of SPC4, having a ribbon or slice obstruction for one of the RBG links while the other can be verified to be ribbon by the agent would be sufficient to establish a counterexample to the conjecture.

References

[1] E. Witten, "Quantum field theory and the Jones polynomial," *Commun. Math. Phys.* **121**, 351–399, 1989.
[2] P. Kucharski, M. Reineke, M. Stosic, and P. Sułkowski, "Knots-quivers correspondence," *Adv. Theor. Math. Phys.* **23**, no. 7, 1849–1902, 2019 [1707.04017].

[6]We formulated the second problem as a computer vision problem, but in the end the goal is also to find actions that lead to the unknot, i.e., the empty braid word, if we allow simplification as well as band moves.

[3] A. Grassi, J. Halverson, F. Ruehle, and J. L. Shaneson, "Dualities of deformed $\mathcal{N} = 2$ SCFTs from link monodromy on D3-brane states," *JHEP* **09**, 135, 2017 [1611.01154].

[4] S. Gukov, J. Halverson, F. Ruehle, and P. Sułkowski, "Learning to unknot," *Mach. Learn. Sci. Tech.* **2**, no. 2, 025035, 2021 [2010.16263].

[5] J. W. Alexander, "A lemma on systems of knotted curves," *Proc. Natl. Acad. Sci. USA* **9**, no. 3, 93–95, 1923.

[6] D. Bar-Natan, I. Bar-Natan, I. Halacheva, and N. Scherich, "Yarn ball knots and faster computations," 2021. *arXiv e-prints* arXiv:2108.10923 [2108.10923].

[7] F. Ruehle, "Data science applications to string theory," *Phys. Rept.* **839**, 1–117, 2020.

[8] M. Culler, N. M. Dunfield, M. Goerner, and J. R. Weeks, "SnapPy, a computer program for studying the geometry and topology of 3-manifolds." Available at http://snappy.computop.org.

[9] N. Kitaev, L. Kaiser, and A. Levskaya, "Reformer: The efficient transformer," in *International Conference on Learning Representations*, 2020.

[10] L. H. Kauffman and S. Lambropoulou, "Hard unknots and collapsing tangles," in *Introductory Lectures on Knot Theory*, Series on Knots and Everything, pp. 187–247, 2011.

[11] B. A. Burton, H.-C. Chang, M. Löffler, C. Maria, A. de Mesmay, S. Schleimer, E. Sedgwick, and J. Spreer, "Hard diagrams of the unknot," *Exp. Math.*, 1–19, 2023. https://doi.org/10.1080/10586458.2022.2161676.

[12] I. A. Dynnikov, "Arc-presentations of links: Monotonic simplification," *Fundam. Math.* **190**, 29–76, 2006.

[13] L. H. Kauffman, N. E. Russkikh, and I. A. Taimanov, "Rectangular knot diagrams classification with deep learning," *J. Knot Theory Its Ramif.* **31**, no. 11, 2250067, 2022.

[14] S. Gukov, J. Halverson, C. Manolescu, and F. Ruehle, "Searching for ribbons with machine learning" [2304.09304].

[15] R. H. Fox, "Some problems in knot theory," in *Topology of 3-Manifolds and Related Topics, Proceedings of the University of Georgia Institute, 1961*, pp. 168–176. Prentice-Hall, Englewood Cliffs, N.J., 1962.

[16] K. Reidemeister, "Elementare Begründung der Knotentheorie," *Abh. Math. Sem. Univ. Hamburg* **5**, no. 1, 24–32, 1927.

[17] J. Hass, J. C. Lagarias, and N. Pippenger, "The computational complexity of knot and link problems," *J. ACM* **46**, 185–211, 1999.

[18] M. Lackenby, "The efficient certification of knottedness and thurston norm," *Adv. Math.* **387**, 107796, 2021.

[19] A. Vaswani, N. Shazeer, N. Parmar, J. Uszkoreit, L. Jones, A. N. Gomez, L. u. Kaiser, and I. Polosukhin, "Attention is all you need," in I. Guyon, U. V. Luxburg, S. Bengio, H. Wallach, R. Fergus, S. Vishwanathan, and R. Garnett (eds.), *Advances in Neural Information Processing Systems*, Vol. 30. Curran Associates, Inc., 2017.

[20] P. S. Novikov, "Über die algorithmische unentscheidbarkeit des wortproblems in der gruppentheorie," *Tr. Mat. Inst. Steklova* **44**, 140, 1955.

[21] Y. Fujita, P. Nagarajan, T. Kataoka, and T. Ishikawa, "Chainerrl: A deep reinforcement learning library," *J. Mach. Learn. Res.* **22**, no. 77, 1–14, 2021.
[22] G. Perelman, "The entropy formula for the ricci flow and its geometric applications," 2002, ar-Xiv:math.DG/0211159v1.
[23] G. Perelman, "Ricci flow with surgery on three-manifolds," 2003, arXiv:math/0303109.
[24] C. Manolescu and L. Piccirillo, "From zero surgeries to candidates for exotic definite four-manifolds" [2102.04391].
[25] J. Rasmussen, "Khovanov homology and the slice genus," *Invent. Math.* **182**, no. 2, 419–447, 2010.
[26] P. Ozsváth and Z. Szabó, "Knot Floer homology and the four-ball genus," *Geom. Topol.* **7**, no. 2, 615–639, 2003.
[27] R. Fox and J. Milnor, "Singularities of 2-spheres in 4-space and cobordism of knots," *Osaka J. Math.* **3**, 257–267, 1966.
[28] C. Herald, P. Kirk, and C. Livingston, "Metabelian representations, twisted Alexander polynomials, knot slicing, and mutation," *Math. Z.* **265**, no. 4, 925–949, 2010.
[29] S. Kinoshita and H. Terasaka, "On unions of knots," *Osaka Math. J.* **9**, 131–153, 1957.
[30] K. He, X. Zhang, S. Ren, and J. Sun, "Deep residual learning for image recognition," in *2016 IEEE Conference on Computer Vision and Pattern Recognition (CVPR)*, pp. 770–778, 2016.
[31] Y. Lecun, L. Bottou, Y. Bengio, and P. Haffner, "Gradient-based learning applied to document recognition," *Proc. IEEE* **86**, no. 11, 2278–2324, 1998.
[32] J. Snoek, H. Larochelle, and R. P. Adams, "Practical bayesian optimization of machine learning algorithms," in F. Pereira, C. Burges, L. Bottou, and K. Weinberger (eds.), *Advances in Neural Information Processing Systems*, Vol. 25. Curran Associates, Inc., 2012.
[33] F. Nogueira, "Bayesian optimization: Open source constrained global optimization tool for python," 2014. https://github.com/fmfn/Bayesian Optimization.
[34] C. Manolescu and L. Piccirillo, "From zero surgeries to candidates for exotic definite four-manifolds" [2102.04391].
[35] R. E. Gompf, M. Scharlemann, and A. Thompson, "Fibered knots and potential counterexamples to the property 2r and slice-ribbon conjectures," *Geom. Topol.* **14**, no. 4, 2305–2347, 2010.
[36] B. Owens and F. Swenton, "An algorithm to find ribbon disks for alternating knots" [2102.11778].
[37] F. Ruehle, "An algorithm for finding ribbon bands," 2023. https://github.com/ruehlef/ribbon.

INDEX

www.ingramcontent.com/pod-product-compliance
Lightning Source LLC
LaVergne TN
LVHW051055210926
848486LV00002B/118
* 9 7 8 1 8 0 7 2 9 0 0 0 9 *